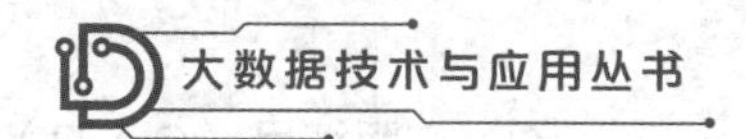

大数据处理技术

基础与应用（Hadoop+Spark）

许桂秋　孙海民　胡贵恒◎主　编
蒋丽丽　肖皇培　陆建波◎副主编

人民邮电出版社
北　京

图书在版编目（CIP）数据

大数据处理技术基础与应用 ：Hadoop+Spark / 许桂秋，孙海民，胡贵恒主编. -- 北京 ：人民邮电出版社，2024.2
（大数据技术与应用丛书）
ISBN 978-7-115-63768-0

Ⅰ. ①大… Ⅱ. ①许… ②孙… ③胡… Ⅲ. ①数据处理软件 Ⅳ. ①TP274

中国国家版本馆CIP数据核字(2024)第023648号

内容提要

本书是一本介绍大数据处理技术的专业图书，力求提高读者对大数据处理的认知水平和动手能力。本书首先介绍大数据技术的相关概念和发展历程，从实践的角度介绍 Hadoop 和 Spark 的安装部署、编程基础和使用方法；然后结合具体案例，重点介绍 Spark RDD、Spark SQL、Spark Streaming、Spark Graph Frame 等的应用思路和方法，并通过具体代码，让读者更好地感受大数据处理技术的效果。

本书既可以作为高等院校计算机、大数据等相关专业的教材，也可以作为大数据技术相关从业人员的参考书，还可作为零基础人员学习 Hadoop 和 Spark 技术的入门图书。

◆ 主　　编　许桂秋　孙海民　胡贵恒
　副 主 编　蒋丽丽　肖皇培　陆建波
　责任编辑　张晓芬
　责任印制　马振武
◆ 人民邮电出版社出版发行　　北京市丰台区成寿寺路 11 号
　邮编　100164　　电子邮件　315@ptpress.com.cn
　网址　https://www.ptpress.com.cn
　山东华立印务有限公司印刷
◆ 开本：787×1092　1/16
　印张：15.5　　　　　　2024 年 2 月第 1 版
　字数：358 千字　　　　2024 年 2 月山东第 1 次印刷

定价：69.80 元

读者服务热线：(010)81055493　印装质量热线：(010)81055316
反盗版热线：(010)81055315

前言

随着物联网和云计算技术的快速发展，人们的生活变得越来越便利，由此积累了种类繁多、体量巨大的数据。这些数据存在于我们生活的每个角落，反映着真实的世界，因此人们希望挖掘出数据背后蕴藏的巨大价值。

Hadoop 和 Spark 是大数据处理技术中的佼佼者，是处理海量数据的理想工具。

本书是大数据处理技术的专业图书，采用理论与实践相结合的方式，由浅入深地介绍大数据处理技术的基本概念和基础知识，并结合实践案例，帮助读者运用所学知识解决现实中的问题。

全书共 11 章，可分为两个部分。

第一部分是 Hadoop 应用，包括第 1～5 章。第 1 章阐述大数据的发展和流程、主流软件和编程语言；第 2 章详细介绍 Hadoop 部署安装与使用；第 3 章介绍 HDFS 基本操作；第 4 章介绍 MapReduce 基本原理与编程实现；第 5 章介绍 Hive 部署与编程基础。

第二部分是 Spark 应用，包括第 6～11 章。第 6 章详细介绍 Spark 部署与编程基础；第 7 章介绍 Spark 的核心数据结构 RDD，为后续其他组件的学习打下基础；第 8 章介绍利用 Spark SQL 处理结构化数据文件；第 9 章介绍利用 Spark Streaming 处理实时数据；第 10 章介绍利用 Spark GraphFrames 处理图计算；第 11 章介绍大数据生态常用工具——Flume、Kafka、Sqoop，并结合实际数据处理流程的案例来加深对其的使用。

本书既可以作为高等院校计算机、大数据等相关专业的教材，也可作为大数据技术相关从业人员的参考书，还可以作为零基础人员学习 Hadoop 和 Spark 技术的入门图书。

由于编者水平有限，书中难免存在一些疏漏和不足之处，恳请广大读者批评指正。

编者

2024 年 2 月

目录

第1章 大数据技术概述

大数据是当今时代的技术热点，无论是传统领域还是新兴领域，到处都有大数据的身影。大数据已经成为我们日常生活中不可缺少的部分。

【学习目标】

1．理解大数据的基本概念。

2．了解大数据的关键技术。

3．熟悉 Hadoop 及其生态系统。

4．熟悉 Spark 及其生态系统。

5．熟悉 Flink 及其生态系统。

6．了解大数据技术使用的编程语言。

1.1 大数据技术简介

大数据时代已经到来，大数据在金融、汽车、零售、政务等领域得到了广泛应用。

1.1.1 大数据技术的发展

1．大数据的产生

1946 年，世界上第一台电子计算机诞生，这时的数据与应用被紧紧地捆绑在文件中。20 世纪 60 年代，信息系统的规模变大、复杂度变高，数据与应用分离的需求开始产生。之后数据库技术开始萌芽，继而蓬勃发展，并在 1990 年后出现了以关系数据库为主导的局面，数据管理技术在 2001 年之前的发展如图 1-1 所示。

从 2001 年开始，互联网迅速发展，数据量成倍递增。据统计，目前，超过 150 亿台（个）设备连接到互联网，全球每秒发送 290 万封电子邮件，每天有累计 2.88 万小时的视频上传到 YouTube 网站。Facebook 网站每日评论达 32 亿条，每天上传照片近 3 亿张，每月处理数据总量约 130 万 TB。意大利 PXR 研究机构数据统计，全球范围内创建、捕获、复制和消费的数据/信息量从 2010 年的 2 ZB 增长到 2020 年的 64.2 ZB。预计到 2025 年，全球数据总量将超过 181 ZB，如图 1-2 所示。

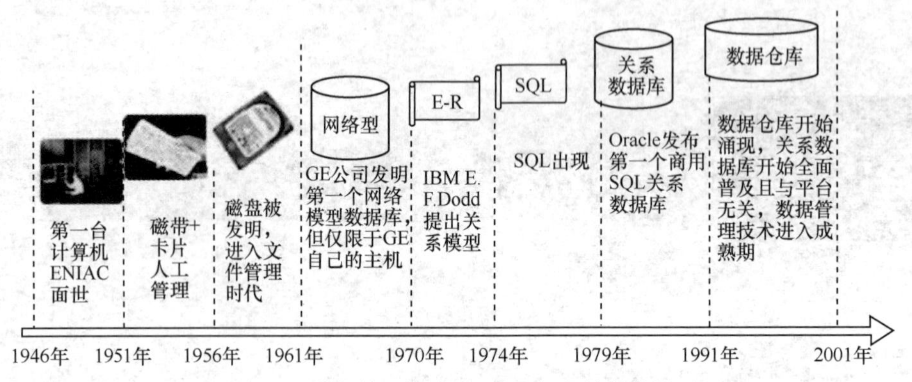

图 1-1　数据管理技术在 2001 年之前的发展

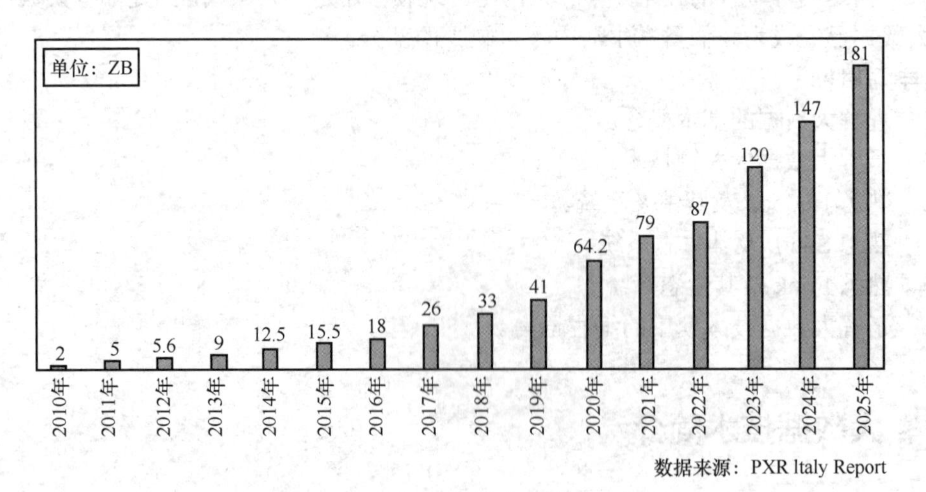

图 1-2　意大利 PXR《大数据产业》报告

2011 年 5 月，EMC World 2011 大会的主题是“云计算相遇大数据”。这次会议除了聚焦 EMC 公司一直倡导的云计算概念，还抛出了“大数据”的概念。2011 年 6 月底，IBM 公司、麦肯锡等众多国外机构发布“大数据”相关研究报告，并予以积极跟进。

2015 年，十八届五中全会首次提出“国家大数据战略”。同年《促进大数据发展行动纲要》发布。

2016 年，《政务信息资源共享管理暂行办法》发布。

2017 年，《大数据产业发展规划（2016—2020 年）》实施。

2020 年 10 月，十九届五中全会发布《中共中央关于制定国民经济和社会发展第十四个五年规划和二〇三五年远景目标的建议》。数字经济成为国家产业结构升级的重要突破点之一。

2021 年 11 月 15 日，工业和信息化部发布《“十四五”大数据产业发展规划》。

2022 年 1 月 12 日，国家发展和改革委员会发布《“十四五”数字经济发展规划》。

2023 年，互联网数据中心（Internet Data Center，IDC）发布的 Global DataSphere 2023 显示，中国数据量规模将从 2022 的 23.88 ZB 增长至 2027 年的 76.6 ZB，年均增长速度达到 26.3%，为全球第一。IDC 中国数据量规模预测如图 1-3 所示。

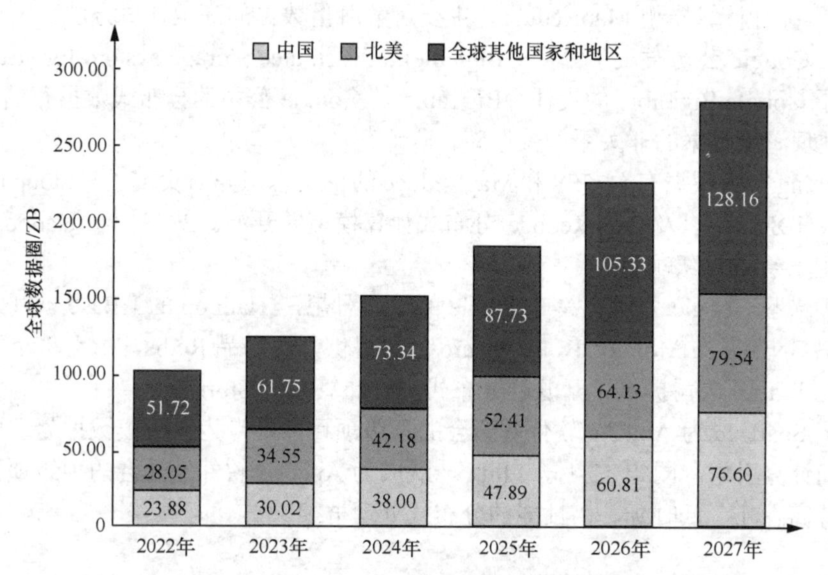

数据来源：IDC Global DataSphere, 2023。IDC将每年被创建、采集或复制的数据集合定义为数据圈。

图 1-3　IDC 中国数据量规模预测

2. 大数据特征

大数据，简单地说，是指无法在一定时间内用常规软件（工具）对其内容进行抓取、管理和处理的数据集合。IBM 公司将大数据的特点总结为 4V，即 Volume（数据量大）、Variety（数据多样性）、Velocity（数据的产生和处理速度快）、Value（价值密度低）。

- 数据量大：大数据的数据量从太字节（TB）级别跃升到拍字节（PB）级别。
- 数据多样性：大数据包括网络日志、视频、图片、地理位置信息等结构化、半结构化、非结构化数据。
- 数据的产生和处理速度快：要求在很短的时间内给出分析结果，否则结果可能会失去价值。
- 价值密度低：以视频为例，在连续不间断的监控过程中，产生的视频数据可能仅一两秒有用。

3. 大数据技术的发展历程

随着应用数据规模的急剧增加，传统系统面临严峻挑战，难以提供足够的存储和计算资源进行数据处理。大数据技术是从各种类型的海量数据中快速获得有价值信息的技术。大数据技术要面对的基本问题，也是核心的问题，就是海量数据如何可靠存储和高效计算。

围绕大数据的核心问题，下面列出了大数据相关技术的发展历程：

2003 年，Google 公司发表了论文“The Google file system”，介绍分布式文件系统（Distributed File System，GFS），主要讲解海量数据的可靠存储方法。

2004 年，Google 公司发表了论文“MapReduce: simplified data processing on large clusters”，介绍并行计算模型 MapReduce，主要讲解海量数据的高效计算方法。

2006 年，Google 公司发表了论文“Bigtable: a distributed storage system for structured data”，介绍了 Google BigTable 的设计。BigTable 是 Google 公司的分布式数据存储系统，是一种用来处理海量数据的非关系型数据库。

Doug Cutting 等人根据有关 GFS 和 MapReduce 的论文思想先后实现了 Hadoop 的分布式文件系统（HDFS），以及 MapReduce 分布式计算模型并开源。2008 年，Hadoop 成为 Apache 软件基金会的顶级项目。

2010 年，借鉴 Google 公司有关 BigTable 的论文思想， Hadoop 的 HBase 被开发并开源。开源组织 GNU 发布 MongoDB，VMware 公司提供开源产品 Redis。

2011 年，Twitter 公司提供了开源的分布式实时计算系统 Storm。

2014 年，Spark 成为 Apache 软件基金会的顶级项目，是专为大规模数据处理而设计的快速通用的计算引擎。同年 12 月，Flink 一跃成为 Apache 软件基金会的顶级项目，是能够同时支持高吞吐、低时延、高性能的分布式处理框架。

1.1.2 大数据的基本处理流程

大数据的基本处理流程主要包括数据采集与预处理、数据存储与管理、数据处理与分析、数据可视化、数据安全与隐私保护等，如表 1-1 所示。数据可视化有时被视为数据分析的一部分，即可视化分析，因此，数据可视化也可归入数据处理与分析。

表 1-1 大数据的基本处理流程

处理流程	描述
数据采集与预处理	利用抽取–转换–加载（Extraction-Transformation-Loading，ETL）工具将异构数据源中的数据（如关系数据）抽取后进行清洗、转换、集成，并加载到数据仓库或数据集中，成为联机分析处理（On line Analytical Processing，OLAP）、数据挖掘的基础；也可以利用日志采集工具（如 Flume、Kafka 等）把实时采集的数据作为流计算系统的输入，进行实时处理分析
数据存储与管理	利用分布式文件系统、数据仓库、关系数据库、NoSQL 数据库、云数据库等，实现对结构化、半结构化和非结构化数据的存储和管理
数据处理与分析	利用分布式并行编程模型和计算框架，结合机器学习和数据挖掘的相关算法，实现对海量数据的处理和分析
数据可视化	对分析结果进行可视化呈现，帮助人们更好地理解数据、分析数据
数据安全与隐私保护	在从大数据中挖掘潜在商业价值和学术价值的同时，构建隐私数据保护体系和数据安全体系，有效保护个人隐私和数据安全

每种大数据技术都有其适用场景，企业可根据具体应用场景选择合适的大数据计算模式，再根据所选的大数据计算模式选择相应的大数据计算产品。大数据计算模式如表 1-2 所示。

表 1-2 大数据计算模式

模式	适用场景	对应产品
批处理计算	针对大规模数据的批量处理	MapReduce、Spark 等
流计算	针对流数据的实时计算	Storm、Streams、Puma、DStream 等
图计算	针对大规模图结构数据的处理	Pregel、GraphX、Giraph、PowerGraph、Hama 等
查询分析计算	针对大规模数据的存储管理和查询分析	Dremel、Hive、Cassandra、 Impala 等

1.2 大数据的主流处理软件

1.2.1 Hadoop

1. Hadoop 简介

Hadoop 是 Apache 软件基金会旗下的一款开源软件，为用户提供系统底层细节透明的分布式基础架构。Hadoop 基于 Java 语言开发，具有很好的跨平台特性，并且可以部署在低成本的计算机集群中。Hadoop 的核心是 HDFS 和 MapReduce。HDFS 是针对谷歌文件系统（Google File System，GFS）的开源实现，是面向普通硬件环境的分布式文件系统。MapReduce 允许用户在不了解分布式系统底层细节的情况下开发并行应用程序，保证了分布式文件系统上数据分析和处理的高效性。

2. Hadoop 的特性

Hadoop 具有以下特性。

高可靠性。Hadoop 采用冗余数据存储方式，可以在一个副本发生故障时，其他副本能使系统正常对外提供服务。

高效性。Hadoop 采用分布式存储和分布式处理技术来高效地处理拍字节（PB）级数据。

高可扩展性。Hadoop 的节点可以扩展到数千个。

高容错性。Hadoop 能自动保存数据的多个副本，并且能够自动对失败的任务进行重新分配。

成本低。Hadoop 采用低配的计算机集群，让普通用户可以用自己的计算机搭建 Hadoop 运行环境。

在 Linux 环境下运行。Hadoop 基于 Java 语言开发，可以较好地运行在 Linux 环境下。

支持多种编程语言。Hadoop 上的应用程序可以使用其他语言编写，如 C++。

3. Hadoop 的核心

（1）HDFS

HDFS 是分布式计算中数据存储与管理的基础。HDFS 架构如图 1-4 所示。

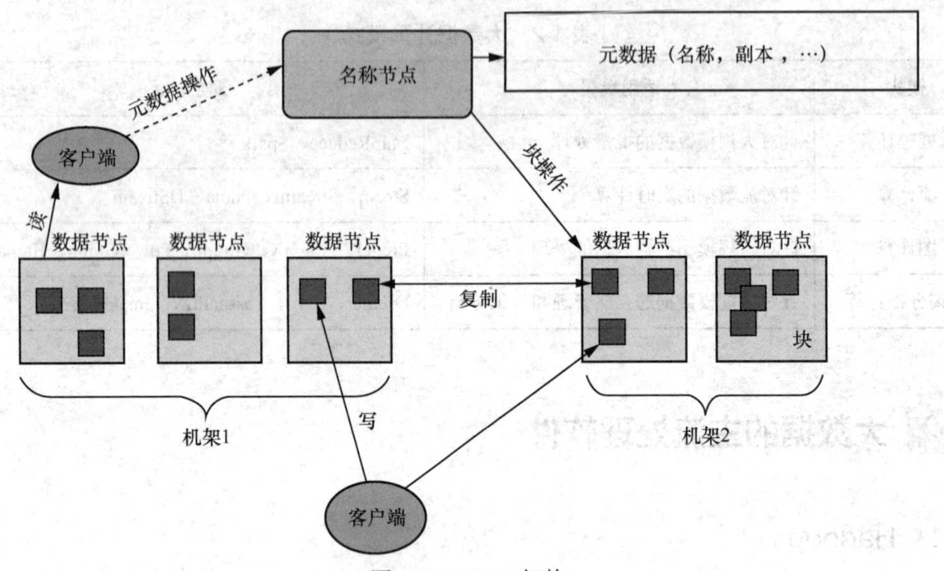

图 1-4　HDFS 架构

HDFS 通过机架感知和副本冗余存储策略来实现数据的存储与管理，如图 1-5 所示。在图 1-5 中，副本 1 保存在机架 1 的主机上，副本 2 保存在机架 2 的主机上，副本 3 保存在机架 2 的另一台主机上。这主要出于对安全和效率的考虑，因为副本 2 的主机如果坏了，那么可以按照就近原则，从同一个机架的其他主机获取副本。

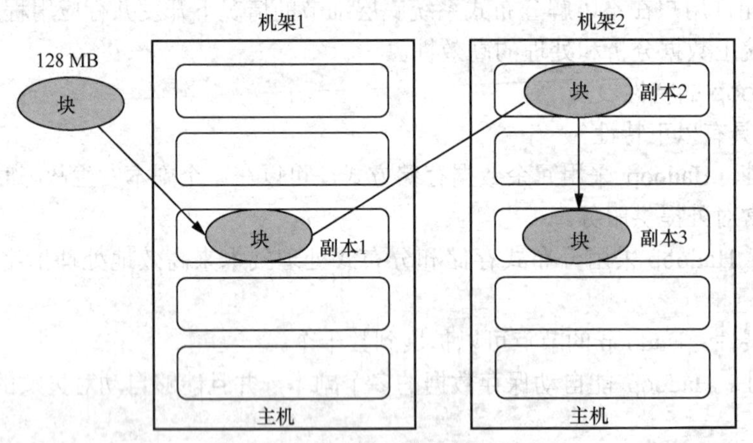

图 1-5　HDFS 数据的存储与管理

（2）MapReduce

MapReduce 是一种编程模型，用于大规模数据集群的并行运算。它将复杂的、运行于大规模集群上的并行计算过程高度地抽象到 Map 和 Reduce 这两个函数上。通俗地说，MapReduce 的核心思想是“分而治之”，如图 1-6 所示。在图 1-6 中，它把输入的大任务（数据集）切分为若干独立的小任务（数据块），分发给主节点所管理的各个分节点来并行完成，之后通过整合各个节点的结果来得到最终结果。

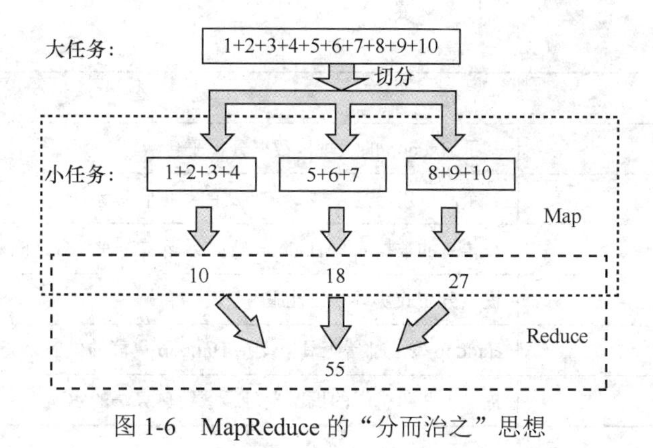

图 1-6　MapReduce 的“分而治之”思想

4. Hadoop 生态系统

狭义的 Hadoop 是一个适合大数据分布式存储和分布式计算的平台，包括 HDFS、MapReduce 和 YARN。

广义的 Hadoop 是一个以 Hadoop 为基础的生态系统，这是一个很庞大的系统，Hadoop 只是其中最重要、最基础的部分。Hadoop 生态系统中的子系统只负责解决某一个特定的问题域，这些子系统不是一个全能系统，而是小而精的多个小系统。Hadoop 生态系统如图 1-7 所示。

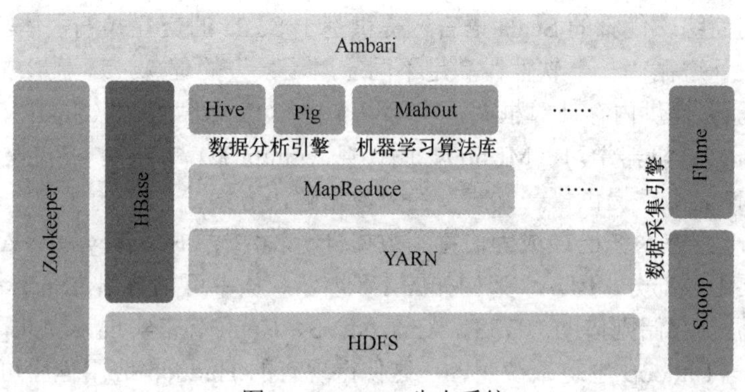

图 1-7　Hadoop 生态系统

Hadoop 生态系统的常用组件及其说明如表 1-3 所示。

表 1-3　Hadoop 生态系统的常用组件及其说明

组件	功能
HDFS	分布式文件系统
YARN	资源调度框架
MapReduce	分布式并行编程模型
HBase	建立在 HDFS 之上的分布式列式数据库
Hive	数据仓库

续表

组件	功能
Pig	基于 Hadoop 的大规模数据分析平台
Flume	一个高可用、高可靠、分布式的数据收集系统
Sqoop	在传统的数据库与 Hadoop 数据存储和处理平台之间进行数据传递的工具
Zookeeper	提供分布式协调一致性服务
Ambari	Hadoop 安装部署工具，支持 Hadoop 集群的供应、管理和监控
Mahout	提供一些可扩展的机器学习领域经典算法的实现

1.2.2 Spark

1．Spark 简介

Spark 最初由美国加利福尼亚大学伯克利分校的 AMP 实验室于 2009 年开发，是一个可应用于大规模数据处理的通用引擎，如今是 Apache 软件基金会下的开源项目之一。Spark 最初的设计目标是使数据分析更快——不仅要快速运行，也要能快速地编写程序，为此 Spark 提供了内存计算，减少了迭代计算时的输入/输出开销。为了使编写程序更容易，Spark 使用简练、优雅的 Scala 语言，提供基于 Scala 的交互式编程体验。

Spark 的设计遵循“一个软件栈满足不同应用场景”的理念，逐渐形成了一套完整的生态系统：既能够提供内存计算框架，也可以支持 SQL 即席查询（Spark SQL）、流计算（Spark Streaming）、机器学习（MLlib）、图计算（GraphX）等应用。Spark 可以部署在资源管理器 YARN 之上，提供“一站式”的大数据解决方案。

如今，Spark 生态系统已经成为伯克利数据分析软件栈（Berkeley Data Analytics Stack，BDAS）的重要组成部分。BDAS 架构如图 1-8 所示，从中可以看出，Spark 专注于数据的处理分析，数据的存储则借助 HDFS、Amazon S3、Ceph 等来实现，因此，Spark 生态系统可以很好地与 Hadoop 生态系统兼容，这使得现有的 Hadoop 应用程序可以非常容易地迁移到 Spark 系统中。

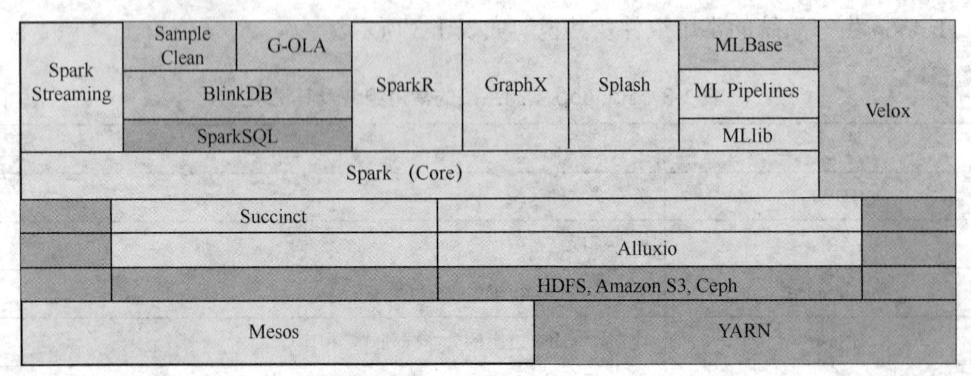

图 1-8　DBAS 架构

2. Spark 的特点

Spark 有以下特点。

运行速度快。Spark 使用先进的有向无环图执行引擎，以支持循环数据流与内存计算。

容易使用。Spark 支持使用 Scala、Java、Python、R 等语言进行编程，且 API 的设计很简洁。

Spark 有助于用户轻松构建并行程序，并且可以通过 Spark Shell 进行交互式编程。

通用性。Spark 提供了完整而强大的技术栈，其中包括 Spark SQL、Spark Streaming、MLlib 和 GraphX。

算法组件可以无缝地被整合在同一个应用中，能够应对复杂的计算。

运行模式多样。Spark 可运行在独立的集群模式上，也可运行在 Hadoop 上，还可运行在 Amazon EC2 等云环境上，并且可以访问 HDFS、Cassandra、HBase、Hive 等多种文件系统。

3. Spark 生态系统

Spark 生态系统以 Spark Core 为核心，利用 YARN、Mesos 等资源调度管理工具来完成应用程序分析与处理，这些应用程序来自 Spark 的不同组件，例如 Spark Streaming、Spark SQL、MLlib、GraphX 等。Spark 生态系统如图 1-9 所示。

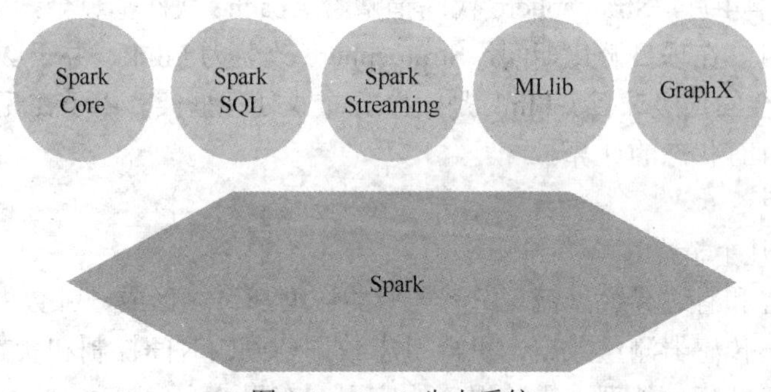

图 1-9　Spark 生态系统

（1）Spark Core

Spark Core 提供 Spark 最基础与最核心的功能，例如内存计算、任务调度、部署模式、故障恢复、存储管理等，主要面向批数据处理。Spark 建立在统一的弹性分布式数据集（Resilient Distributed Dataset，RDD）之上，可以以基本一致的方式应用于不同的大数据处理场景。

（2）Spark Streaming

Spark Streaming 支持高吞吐量、可容错处理的实时流数据处理，其核心思路是将流数据分解成一系列短小的批处理作业，每个批处理作业可以使用 Spark Core 进行快速处理。Spark Streaming 支持从 Kafka、Flume、Twitter、ZeroMQ、Kinesis、TCP Socket 等数据源获取数据，并且可以使用复杂的算法和高级功能进行数据处理。处理后的数据可以被推送到文件系统或数据库。

（3）Spark SQL

Spark SQL 是一种结构化的数据处理模块。它提供的一个称为 DataFrame 的编程抽象可以作为分布式 SQL 查询引擎。

一个 DataFrame 相当于一个列数据的分布式采集组织，其形式类似于关系数据库中的表。它可以利用多种方式进行构建，例如结构化数据文件、Hive、外部数据库或 RDD。

（4）MLlib

MLlib 提供了常用机器学习算法的实现方式，其中包括聚类、分类、回归分析、协同过滤等算法，降低了机器学习的门槛。开发人员只要具备一定的理论知识，就能使用 MLlib 进行机器学习的相关任务。

（5）GraphX

GraphX 是 Spark 上的分布式图形处理架构，可用于图计算。

1.2.3 Flink

Flink 起源于由柏林工业大学、柏林洪堡大学和哈索・普拉特纳研究院联合开展的“Stratosphere: Information Management on the Cloud”项目，该项目在当时的社区具有一定的知名度。2014 年 4 月，Stratosphere 代码捐赠给 Apache 软件基金会，成为 Apache 软件基金会孵化器项目。在项目孵化期间，Stratosphere 改名为 Flink，并于 2014 年 12 月成为 Apache 软件基金会的顶级项目。Flink 是一个分布式处理框架，用于在无边界数据流和有边界数据流上进行有状态的计算。

1. Flink 特点

Flink 有以下几个特点。

- 处理无边界和有边界数据流。任何类型的数据都可以形成一种数据流，例如信用卡交易数据、传感器测量数据、机器日志数据精确的时间控制和状态化使 Flink 的运行时能够运行任何处理无界数据流的应用。有界流则由一些 Flink 专为固定大小数据集特殊设计的算法和数据结构进行内部处理。
- 将应用部署到任意地方。Flink 集成了所有常见的集群资源管理器，例如 Hadoop YARN、Apache Mesos 和 Kubernetes，但也可以作为独立集群运行。
- 运行任意规模应用。Flink 可在任意规模上运行有状态的流式应用，将应用程序并行化为数千个任务，这些任务分布在集群中并发执行。
- 利用内存性能。有状态的 Flink 程序针对本地状态访问进行了优化。任务的状态始终保留在内存中，任务通过访问本地状态（通常在内存中）来进行所有的计算，这使得处理时延非常低。Flink 通过定期和异步方式对本地状态进行持久化存储，来保证故障场景下端到端精确一次和状态一致性。

2. Flink 生态系统

Flink 生态系统如图 1-10 所示。

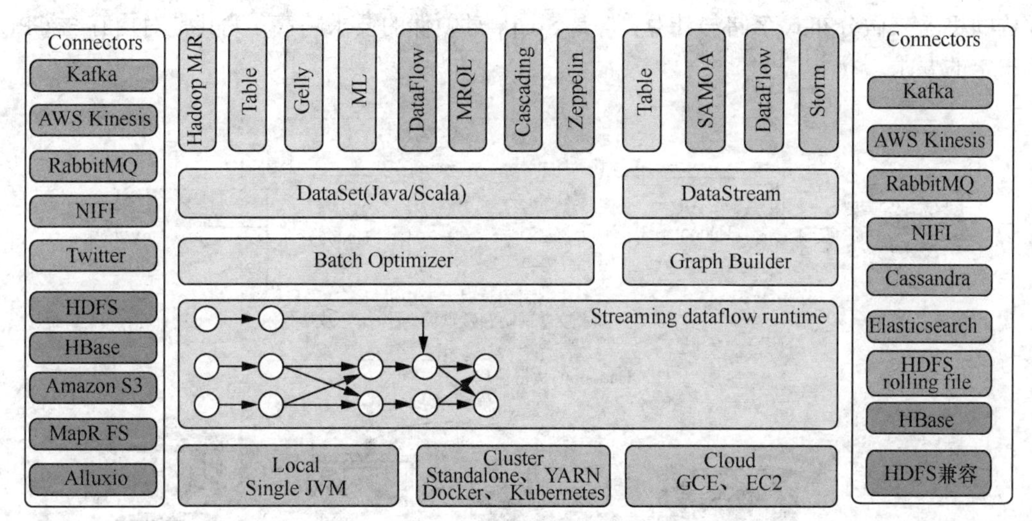

图 1-10　Flink 生态系统

3. Flink 的重要概念

（1）Streams

Stream（流）分为有限数据流与无限数据流。无边界数据流是有始无终的数据流，有边界数据流是限定大小的有始有终的数据流，二者的区别在于无边界数据流的数据会随时间的推演而持续增加，计算持续进行且不存在结束的状态；而有边界数据流的数据大小固定，计算最终会完成并处于结束的状态。

（2）State

State（状态）是计算过程中的数据信息，在容错恢复和 Checkpoint 中有重要的作用，流计算在本质上是增量处理，因此需要不断查询保持状态。

（3）Time

Time 分为事件时间（Event Time）、摄入时间（Ingestion Time）、处理时间（Processing Time）。Flink 的无边界数据流是一个过程持续的数据流，时间是判断业务状态是否滞后、数据处理是否及时的重要依据。

（4）API

API 通常分为 3 层，由上而下可分为 SQL / Table API 层、DataStream API 层、Process Function 层。越接近 SQL / Table API 层，API 的表达能力越弱，抽象能力越强。而 Process Function 层 API 的表达能力非常强，可以进行多种灵活的操作，但抽象能力相对较小。

1.2.4 Hadoop 与 Spark 的对比

Hadoop 与 Spark 的执行流程对比如图 1-11 所示。由图 1-11 可以看到，Spark 最大的特点就是将计算数据和中间结果都存储在内存中，大大减少了输入/输出开销。使用 Hadoop 进行迭代计算非常耗资源，因为每次迭代都需要从磁盘中写入、读取中间数据。Hadoop 与 Spark 在执行逻辑回归时所耗费的时间也相差很大。此外，Hadoop 可以使用低配的、异构的机器来做分布式存储与计算，但是 Spark 对硬件的要求略高，尤其是对内存与 CPU 有一定的要求。

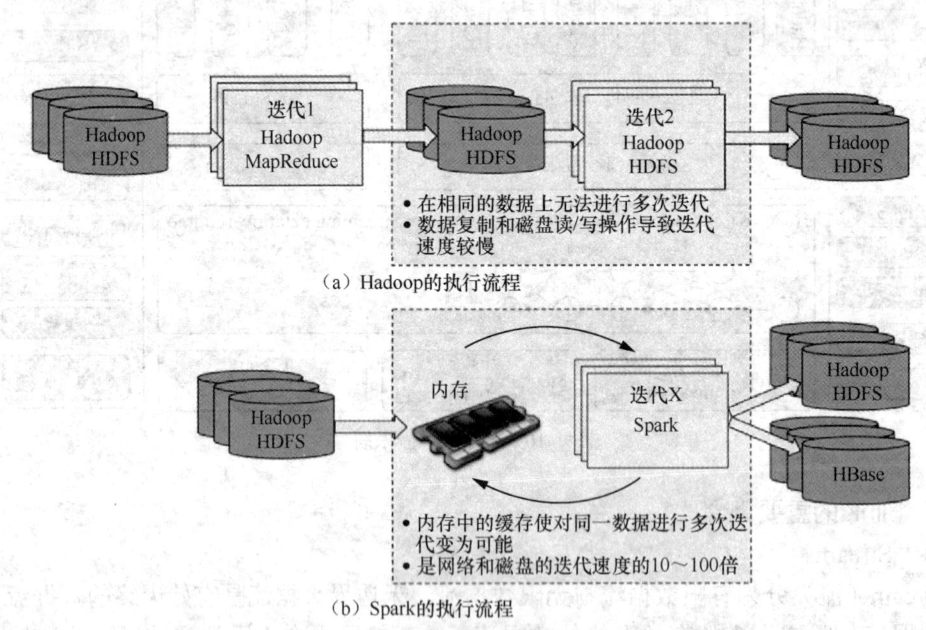

图 1-11　Hadoop 与 Spark 的执行流程对比

尽管相对于 Hadoop 而言，Spark 具有较大优势，但并不能完全替代 Hadoop，只是可以替代 Hadoop 中的 MapReduce 计算模型。实际上，Spark 已经很好地融入 Hadoop 生态系统，成为其中的重要一员。它借助 YARN 实现了资源调度管理，借助 HDFS 实现了分布式存储。

1.3 大数据的主流编程语言

1.3.1　Python 语言

Python 语言诞生于 1991 年，一直是非常流行的计算机语言。Python 可以应用于命令行窗口、图形用户界面（包括 Web 界面）、客户端和服务器端 Web 界面、大型网站后端、云服务（第三方管理服务器）、移动设备、嵌入式设备等计算环境，可实现小到一次性脚本代码，大到几十万行的系统级代码的开发。

Python 是一种面向对象的解释型编程语言，拥有丰富且强大的库，具有简单易学、免费、开源、可移植性强、面向对象、可扩展性强、规范代码等特点。如今，Python 已经成为继 Java 语言和 C++语言之后的第三大编程语言。

以下代码是一段简单的 Python 代码。

```
language = 7
print("Language %s: This is my first Python program." % language)
```

Python 已经和我们的日常用语很接近了，它也因此被称为“伪代码”。虽然 Python 的计算能耗提高了，但运行效率的提高对于编程人员来说是有利的。

1.3.2 Java 语言

Java 是一种面向对象的编程语言，吸收了 C++语言的优点，摒弃了 C++语言中难以理解的多继承、指针等概念，具有功能强大和简单易用这两个特征，允许程序员以优雅的思维方式进行复杂的编程。Hadoop 及其他大数据处理技术也使用 Java 语言进行编程。

JDK(Java Development Kit)称为 Java 开发包或 Java 开发工具，是一个用于编写 Applet 小程序和应用程序的开发环境。JDK 是 Java 的核心，包括 Java 运行环境、Java 工具和 Java 的核心类库（Java API）。无论哪种 Java 应用服务器，其实质都是内置了某个版本的 JDK。主流的 JDK 是 Sun 公司发布的 JDK。

另外，Java API 类库中的 JavaSE API 子集和 Java 虚拟机两部分也称为 Java 运行环境（Java Runtime Environment，JRE），JRE 是支持 Java 程序运行的标准环境。

JRE 是一个运行环境，JDK 是一个开发工具，因此编写 Java 程序时需要 JDK，而运行 Java 程序时需要 JRE。JDK 已经包含 JRE，因此，只要安装了 JDK，就可以编写 Java 程序，也可以正常运行 Java 程序。但是，JDK 包含了许多与运行无关的内容，占用的空间较大，因此若运行普通的 Java 程序，则无须安装 JDK，只安装 JRE 即可。

1.3.3 Scala 语言

Scala 是一种现代的多范式编程语言，平滑地集成了面向对象和函数式语言的特性，旨在以简练、优雅的方式来表达常用的编程模式。Scala 语言的名称来自“可扩展的语言（A Scalable Language）”，从写一个小脚本到建立一个大系统，Scala 语言均可胜任。Scala 运行在 Java 虚拟机（Java Virtual Machine，JVM）上，并兼容现有的 Java 程序。

Spark 设计目的之一是使程序编写得更快、更容易，这也是 Spark 选择 Scala 的原因所在。总体而言，Scala 具有以下突出的优点。

- Scala 具备强大的并发性，支持函数式编程，可以更好地支持分布式系统。
- Scala 语法简洁，能提供优雅的 API。
- Scala 运行速度快，而且能融合到 Hadoop 生态系统中。

1.4 本章小结

大数据技术具有综合性强的知识体系，在学习该技术之前，读者有必要先建立对大数据技术体系的整体性认识。因此，本章主要介绍了大数据技术的发展和流程、主流的软件和语言。此外，本书配套的资源可以帮助读者更加有效地理解本章内容。

第 2 章　Hadoop 部署与使用

Hadoop 的安装有 3 种模式：单机模式、伪分布式模式、完全分布式模式。这 3 种模式的特点和区别如下。

单机模式是指 Hadoop 运行在一台主机上，按默认配置以非分布式模式运行一个独立的 Java 进程。单机模式的特点是没有分布式文件系统，直接在本地操作系统的文件系统上读/写数据，不需要加载任何 Hadoop 守护进程。单机模式一般用于本地 MapReduce 程序的调试，是 Hadoop 的默认模式。

伪分布式模式是指 Hadoop 运行在一台主机上，使用多个 Java 进程来模拟完全分布式模式的各类节点。伪分布式模式具备完全分布式的所有功能，常用于调试程序。

完全分布式模式也叫集群模式，是指 Hadoop 运行在多台主机上，各台主机按照相关配置运行相应的 Hadoop 守护进程。完全分布式模式是真正的分布式环境，可用于实际的生产环境中。

本章将介绍 Hadoop 完全分布式模式的安装方法。

【学习目标】

1．熟悉 Linux 操作系统，掌握常见命令的使用方法。
2．掌握 Hadoop 完全分布式集群的构建与配置方法。
3．掌握 Hadoop 集群基本信息的查询方法。

2.1 Linux 基本操作

【任务描述】创建用户 user01，该用户对应一个主目录/home/user01。之后，在主目录下新建一个文件 demo02.txt，并修改 user01 用户对该文件的权限。

2.1.1 Linux 简介

Linux 操作系统是一个诞生于网络、成长于网络且成熟于网络的操作系统。1991 年，芬兰赫尔辛基大学的学生 Linus Torvalds 写出了属于自己的 Linux 操作系统，该系统的版本为 Linux 0.01。之后，他将 Linux 操作系统放在网络上，供大家下载使用。一大批开发人员加入了对 Linux 操作系统开发和完善的过程中，这让 Linux 操作系统逐渐成长并壮大起来。

Linux 操作系统核心版本编号的注解如图 2-1 所示。Linux 3.0 以后的版本都遵循该标准，并且后续的版本都是在前序版本的基础上经过完善开发出来的。

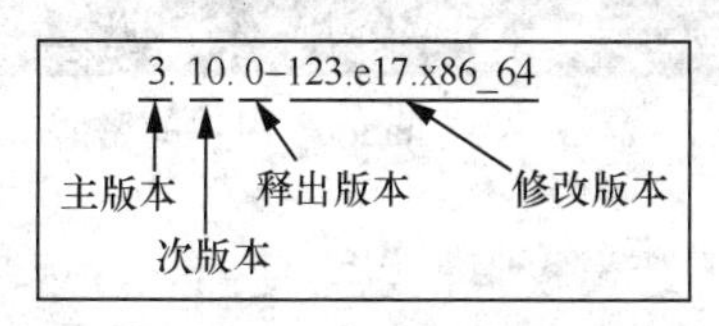

图 2-1　Linux 核心版本编号的注解

Linux 操作系统的核心是 Linux 内核。Linux 操作系统发行版是专门为使用者量身打造的“Linux 内核 + 软件 + 工具 + 文档 + 可完全安装程序”套件的综合发布版本，有利于开发人员在 Linux 操作系统上完成工作。Linux 内核和 Linux 发行版的关系如图 2-2 所示。

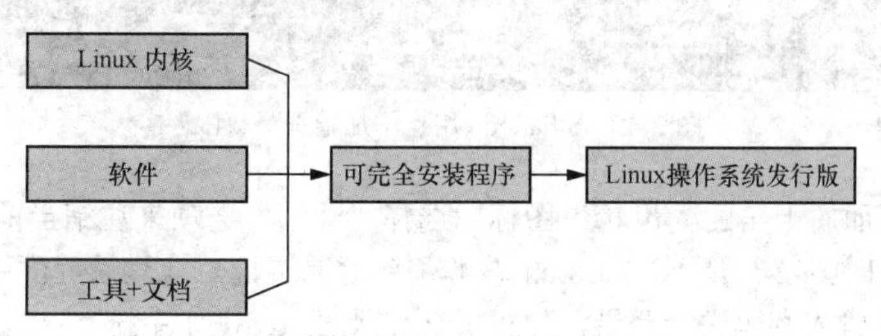

图 2-2　Linux 内核和 Linux 操作系统发行版的关系

目前，Linux 操作系统发行版主要分为两大类系统：一种是使用红帽软件包管理器方式安装软件的系统，主要包括 RedHat、Fedora、SUSE 等；另一种是使用 dpkg 软件包管理器方式安装软件的系统，包括 Debian、Ubuntu、B2D 等。

目前，Linux 操作系统的主要应用场景如下。对于企业环境，Linux 操作系统可应用于网络服务器、关键任务的应用（金融数据库、大型企业网管环境）、学术机构的高效能运算任务等场景。对于个人环境，Linux 操作系统可用于桌面计算机（桌面操作系统）、手持系统（移动端系统）、嵌入式系统（如路由器、防火墙等）应用。Linux 还可用于云程序（云端虚拟机资源）、端设备等应用。

2.1.2　新建与删除用户

本书实验环境已创建一个名称为 ubuntu 的普通用户，后续所有操作会使用该用户名登录 Linux 操作系统。

首先，我们使用 ubuntu 用户登录 Linux 操作系统（sudo 命令需要输入 ubuntu 用户的密码），然后打开一个终端，使用如下命令创建一个用户 hadoop。

```
$sudo  useradd  - m  hadoop  - s  /bin/bash
```

这条命令创建了一个可以登录 Linux 操作系统的 hadoop 用户，并使用 /bin/bash 作为 Shell。我们使用如下命令为 hadoop 用户设置密码。

```
$sudo  passwd  Hadoop
```

我们把密码设置为 123456，并按照提示输入两次密码，如图 2-3 所示。

```
ubuntu@hadoop-node:~$ sudo passwd hadoop
输入新的 UNIX 密码:
重新输入新的 UNIX 密码:
passwd：已成功更新密码
```

图 2-3　设置 hadoop 用户的密码

我们为 hadoop 用户增加管理员权限，以方便部署，规避一些对于初学者来说比较棘手的权限问题，方便复现实验内容。具体命令如下。

```
$sudo  adduser  hadoop  sudo
```

得到的提示如图 2-4 所示。

```
ubuntu@hadoop-node:~$ sudo adduser hadoop sudo
正在添加用户"hadoop"到"sudo"组...
正在将用户“hadoop”加入到“sudo”组中
完成。
```

图 2-4　为 hadoop 用户增加管理员权限

单击操作界面右上角的“齿轮”图标，选择“注销”选项来注销当前登录的 ubuntu 用户，这时会返回图 2-5 所示的 Linux 操作系统登录界面。在登录界面中选择刚创建的 hadoop 用户，并输入密码进行登录，便可成功登录 Linux 操作系统。

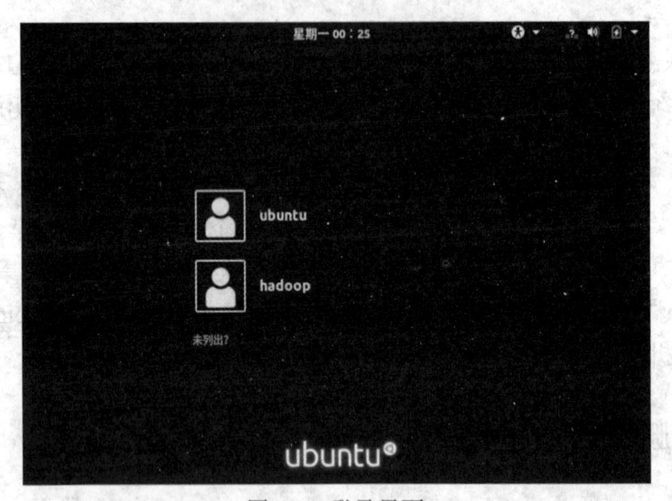

图 2-5　登录界面

我们使用 ubuntu 用户进行登录，使用如下命令删除 hadoop 用户。

```
$sudo  userdel  Hadoop
```

2.1.3　目录权限的查看与修改

Linux 操作系统的目录权限可以通过命令来查看及修改。

1. 查看目录权限

在Ubuntu操作系统中，每个文件有读（r）、写（w）、执行（x）这3种权限，对应的权限值分别为4、2、1。拥有者（u）、同组用户（g）、其他用户（o）可对文件拥有不同的权限。文件的具体信息可通过ls - s filename或ll filename命令进行查看，包含7个字段，示例如下。

```
-rw-rw-r--  1    ubuntu   ubuntu   0     Jan 9  21:07  test
字段1       字段2  字段3     字段4   字段5       字段6       字段7
```

① 字段1表示文件类型及权限，包含10个字符。各字符的含义如下。

- 第1个字符表示文件类型，具体取值如下。

b：块设备，是一些提供系统存取数据的接口设备，例如硬盘。块设备可以随机读取字符。

c：字符设备，是一些串行端口的接口设备，例如键盘。字符设备只能顺序读取字符。

d：目录。

l：链接。

s：套接字，用于进程间通信。

p：命名管道先进先出，用于进程间通信。

-：一般文件，可分为文本文件和二进制文件。

- 第2～4个字符表示拥有者（u）的权限。
- 第5～7个字符表示同组用户（g）的权限。
- 第8～10个字符表示其他用户（o）的权限。

② 字段2表示文件的链接数。

③ 字段3表示用户ID。

④ 字段4表示用户组ID。

⑤ 字段5表示文件大小。

⑥ 字段6表示文件创建日期和时间。

⑦ 字段7表示文件名。

2. 修改文件权限

Ubuntu操作系统中使用chmod和chown这两个命令来管理文件权限。

（1）chmod命令

chmod命令用于修改用户的文件权限，包括修改特定用户的特定权限和为所有用户同时指定权限这两种用法，具体如下。

修改特定用户的特定权限：修改指添加、取消等操作，特定用户指拥有者（u）、同组用户（g）、其他用户（o）、所有用户（a），特定权限指读（r）、写（w）、执行（x）。具体示例如下。

```
$chmod  o-w    filename // 取消其他用户的写权限
$chmod  u+wx   filename // 为拥有者添加写和执行权限
$chmod  a+rwx  filename // 为所有用户添加读、写和执行权限
```

为所有用户同时指定权限：权限由3位数字标识，每位数字表示拥有者（u）、同组用户（g）、其他用户（o）的权限；4、2、1依次表示读（r）、写（w）、（x）权限，这些数字可以相加，表示组合权限。文件的读、写、执行权限说明如表2-1所示。具体示例如下。

```
$chmod  777  filcname // 拥有者、同组用户、其他用户都拥有读、写、执行权限
$chmod  600  filename // 拥有者拥有读和写权限，同组用户、其他用户不拥有任何权限
$chmod  421  filename // 拥有者拥有读权限，同组用户拥有写权限，其他用户拥有执行权限
```

表 2-1 文件的读、写、执行权限说明

权限	代码表示（英文单词）	权限值
读	r（read）	4
写	w（write）	2
执行	x（excute）	1

（2）chown 命令

chown 用于修改文件的拥有者和组，主要用法如下。

```
$chown  username[:groupname]  filename
// 修改 filname 的拥有者为 username，用户组为 groupname
```

2.1.4 Linux 的常用命令

1. 关机和重启命令

（1）关机命令

关机的命令如下。

```
$shutdown  -h  now          // 立刻关机
$shutdown  -h  5            // 5 min 后关机
$poweroff                   // 立刻关机
```

（2）重启命令

重启的命令如下。

```
$shutdown  -r  now          // 立刻重启
$shutdown  -r  5            // 5 min 后重启
$reboot                     // 立刻重启
```

2. 帮助命令

（1）--help 命令

--help 命令的示例如下。

```
$shutdown  --help:
$ifconfig  --help: // 查看网络适配器信息
```

（2）man 命令

man 命令用于打开命令说明书，示例如下。

```
$man  shutdown
```

注意，man shutdown 打开命令说明书之后，使用 Q 键退出命令说明书。

3. 目录操作命令

（1）目录切换

目录切换命令语法：cd 目录。具体示例如下。

```
$cd  /              // 切换到根目录
$cd  /usr           // 切换到根目录下的 usr 目录
$cd  ../            // 切换到上一级目录
$cd  ~              // 切换到主目录
$cd  -              // 切换到上次访问的目录
```

（2）目录查看命令

目录查看命令语法：ls [-al]。具体示例如下。

```
$ls            // 查看当前目录下的所有目录和文件
$ls  -a        // 查看当前目录下的所有目录和文件（包括隐藏的文件）
$ls  -l        // 或-ll，列表查看当前目录下的所有目录和文件。列表查看可以显示更多信息
$ls  /dir      // 查看指定目录下的所有目录和文件，例如 ls /usr
```

（3）目录操作

① 创建目录命令语法：mkdir 目录。具体示例如下。

```
$mkdir  aaa           // 在当前目录下创建一个名为 aaa 的目录
$mkdir  /usr/aaa      // 在指定目录下创建一个名为 aaa 的目录
```

② 删除文件或目录命令语法：rm [-rf] 目录。具体示例如下。

删除文件命令的示例如下。

```
$rm  文件             // 删除当前目录下的文件
$rm  -f  文件         // 删除当前目录下的文件（不询问）
```

删除目录命令的示例如下。

```
$rm  -r  aaa          // 递归删除当前目录下的 aaa 目录
$rm  -rf  aaa         // 递归删除当前目录下的 aaa 目录（不询问）
```

删除全部文件和目录命令的示例如下。

```
$rm  -rf  *           // 将当前目录下的所有目录和文件全部删除
$rm  -rf  /*          // 将根目录下的所有文件全部删除。这是“自杀”命令！慎用！慎用！慎用！
```

注意：rm 命令不仅可以删除目录，还可以删除其他文件或压缩包。为了方便大家理解这个命令的用法，我们这样表述：无论删除任何目录或文件，都直接使用语法：rm [-rf] 目录/文件/压缩包。

③ 目录修改命令语法：mv 和 cp。具体如下。

目录重命名命令语法：mv 当前目录新目录。具体示例如下。

```
$mv  aaa  bbb                  // 将当前目录名 aaa 改为新目录名 bbb
```

注意：mv 命令不仅可以对目录进行重命名，还可以对各种文件、压缩包等进行重命名。

剪切目录命令语法：mv 目录名称目录的新位置。具体示例如下。

```
$mv  /usr/tmp/aaa  /usr    // 将/usr/tmp 目录下的 aaa 目录剪切到新位置/usr 目录下
```

注意：mv 命令不仅可以对目录进行剪切操作，还可以对文件、压缩包等执行剪切操作。

拷贝目录命令语法：cp-r 目录名称目录拷贝的目标位置，其中，-r 表示递归。具体示例如下。

```
$cp  /usr/tmp/aaa  /usr  // 将/usr/tmp 目录下的 aaa 目录复制到目标位置/usr 目录下
```

注意：cp 命令不仅可以复制目录，还可以复制文件、压缩包等。拷贝文件和压缩包时不用写 -r 递归，因此上述代码中没有-r。

④ 搜索目录命令语法：find 目录参数文件名称。具体示例如下。

```
$find  /usr/tmp  -name  'a*'
// 查找/usr/tmp 目录下的所有以 a 开头的目录或文件
```

4．文件操作命令

（1）新建文件

新建文件命令语法：touch 文件名。具体示例如下。

```
$touch  aa.txt   // 在当前目录创建一个名为 aa.txt 的文件
```

（2）删除文件

删除文件命令语法：rm -rf 文件名。

（3）修改文件

修改文件命令语法：vi 或者 vim。

vi 基本上可以分为 3 种模式：命令模式（Command Mode）、插入模式（Insert Mode）、底行模式（Last Line Mode）。各模式的功能如下。

① 命令行模式

命令行模式可以控制光标的移动，删除或查找字符，移动或复制某区段，以及进入插入模式或底行模式。

命令行模式常用的命令（键）如下。

控制光标移动：↑，↓，j。

删除当前行：dd。

查找字符：/字符。

进入编辑模式：ioa。

进入底行模式：:。

② 插入模式

只有在插入模式下，才可以进行文字输入。按 ESC 键可回到命令行模式。

③ 底行模式

在底行模式下，我们可以保存文件或退出 vi，也可以设置编辑环境，例如寻找字符串、列出行号等。

底行模式常用的命令（键）如下。

退出编辑状态：:q。

强制退出 vi：:q!。

保存文件并退出 vi：:wq。

（4）文件的查看命令

文件的查看命令有以下几种。

cat 命令可以看文件的最后一屏内容。具体示例如下。

```
$cat  sudo.conf    // 使用 cat 命令查看/etc/sudo.conf 文件，只能显示最后一屏内容
```

more 命令可以显示百分比。在该命令中，可以通过回车键换向下一行，通过空格键换向下一页，通过 Q 键退出查看状态。具体示例如下。

```
$more  sudo.conf  // 使用 more 命令查看/etc/sudo.conf 文件，可以显示百分比
```

less 命令可以翻页查看文件内容。在该命令中，可以使用键盘上的 PageUp 键和 PageDown 键向上和向下翻页，使用 Q 键退出查看状态。具体示例如下。

```
$less  sudo.conf  // 使用 less 命令查看/etc/sudo.conf 文件
```

tail 可以指定行数或者动态查看文件内容。在该命令中，可以使用组合键 Ctrl+C 退出查看状态，具体示例如下。

```
$tail  -10  sudo.conf  // 使用 tail -10 命令查看/etc/sudo.conf 文件的后 10 行内容
```

2.1.5　任务实现

创建一个新用户 user01，他对应主目录/home/user01。之后在主目录下新建一个文件 demo02.txt，并改变用户对该文件的权限。具体操作步骤如下。

步骤 1：打开一个终端（也可使用组合键 Ctrl+Alt+T 打开）。

步骤 2：创建新用户 user01，命令如下。

```
$sudo  useradd  -m  user01  -s  /bin/bash
```

步骤 3：设置新用户 user01 的密码为 123456，命令如下。

```
$sudo  passwd  user01
```

步骤 4：为新用户 user01 添加管理员权限，命令如下。

```
$sudo  adduser  user01  sudo
```

步骤 5：注销当前用户，登录新用户 user01。

步骤 6：打开一个终端。

步骤 7：进入主目录，命令如下。

```
$cd  ~
```

步骤 8：在主目录下新建一个文件 demo02.txt，命令如下。

```
$touch  demo02.txt
```

步骤 9：修改新文件 demo02.txt 的权限，命令如下。

```
$chmod  777  demo02.txt
```

2.2　搭建 Hadoop 完全分布式集群

【任务描述】Hadoop 集群环境分为单机版环境、伪分布式环境和完全分布式环境。本节任务是搭建 Hadoop 完全分布式集群，要求搭建后的集群有 1 个主节点和 2 个子节点。

2.2.1　关闭防火墙

如果不关闭 Ubuntu 操作系统的防火墙，则会出现以下几种情况。

① 无法正常访问 Hadoop HDFS 的 Web 管理页面。

② 会导致后台某些运行脚本（例如后面要学习的 Hive 程序）出现“假死”状态。

③ 在删除和增加节点时，会让数据迁移的处理时间更长，甚至不能正常完成相关操作。

由此可知，我们要部署 Hadoop 完全分布式集群，就要先关闭防火墙。关闭防火墙的命令和得到的结果如下。

```
$sudo  ufw  disable
Firewall stopped and disabled on system startup
```

使用以下命令查看防火墙状态，得到的状态（Status）为不活动（inactive），这说明防火墙已经关闭。

```
$sudo  ufw  status
Status: inactive
```

2.2.2　安装 SSH

SSH 是 Secure Shell 的缩写，其意为安全外壳。它是一种建立在应用层基础上的安全协议，是目前较可靠、专为远程登录会话和其他网络服务提供安全保障的协议。SSH 协议可以有效防止远程管理过程中的信息泄露。

SSH 由客户端软件和服务端软件组成。在安装 SSH 时，Ubuntu 操作系统需要连接互联网。具体步骤如下。

步骤 1：安装 SSH 客户端软件。

Ubuntu 操作系统默认已安装 SSH 客户端软件，我们通过以下命令查看是否已安装。如果返回结果包含“openssh-client”，则说明已经安装 SSH 客户端软件。

```
$sudo  dpkg  -l  |grep  ssh
```

若返回结果不包含“openssh-client”，则说明未安装 SSH 客户端软件，这时使用以下命令进行安装。

```
$sudo  apt-get  install  openssh-client
```

步骤 2：安装 SSH 服务端软件。

Ubuntu 操作系统默认没有安装 SSH 服务端软件，安装 SSH 服务端软件的命令如下。

```
$sudo  apt-get  install  openssh-server
```

重启 SSH 服务，命令如下。

```
$sudo  /etc/init.d/ssh  restart
```

2.2.3　安装 Xshell 及 Xftp（可选）

采用 Xshell 可以通过 SSH 协议远程连接 Linux 主机，采用 Xftp 可实现安全地在 UNIX/Linux 和 Windows 之间传输文件。打开 NetSarang 官网，下载（最新的）Xshell 及 Xftp 免费版本。本书采用的是 Xshell 6.0 及 Xftp 6.0 免费版本。Xshell 和 Xftp 的安装较简单，只需要双击安装文件并默认安装。

安装 Xshell 及 Xftp 后，打开 Xshell，选中界面左侧的所有会话，并单击鼠标右键，在弹出的菜单中选择“新建”→“会话”，如图 2-6 所示。

在“连接”类别中设置名称及主机信息，其中，主机信息是安装 Ubuntu 操作系统的设备的 IP 地址，输入连接信息如图 2-7 所示。

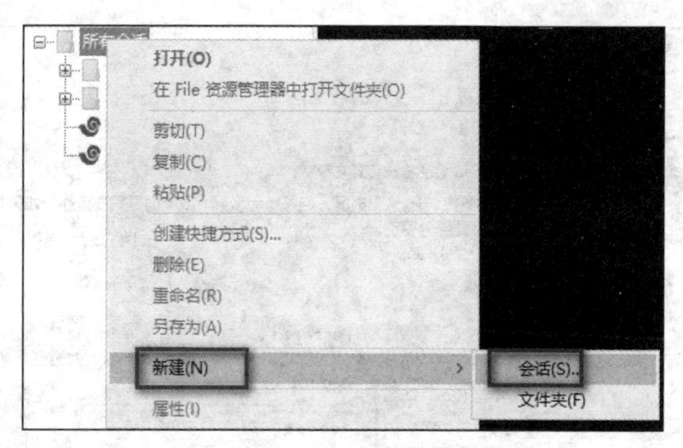

图 2-6　在 Xshell 中新建会话

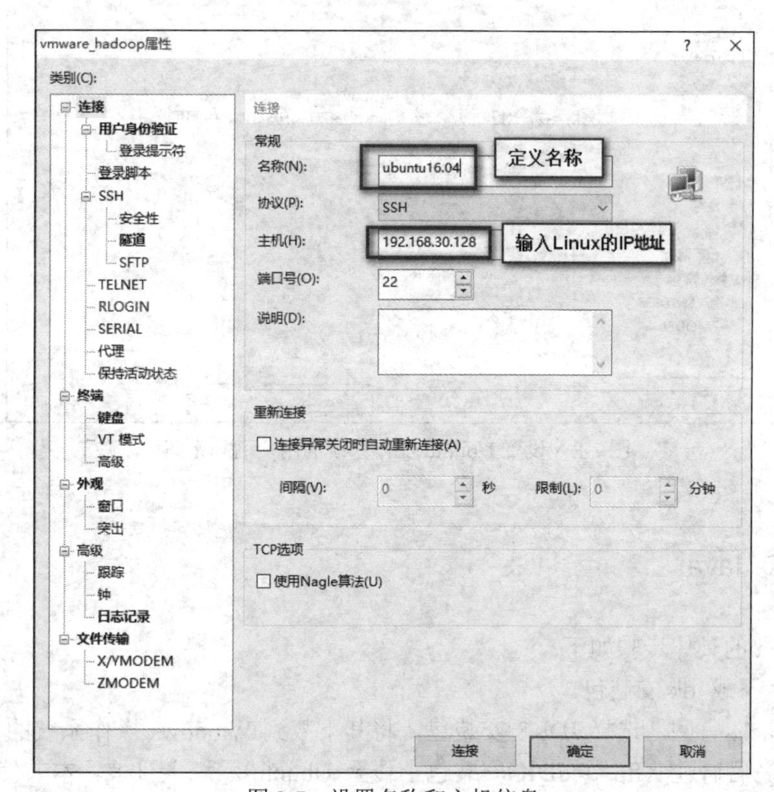

图 2-7　设置名称和主机信息

说明：如果要查看安装 Ubuntu 操作系统的设备的 IP 地址，可使用的命令如下。

```
$ifconfig
```

假设，上述命令运行后得到以下结果，这表示该设备的 IP 地址是“192.168.30.128”。这里的 IP 地址是自动获取的，建议读者将 IP 地址设置为静态 IP 地址。

```
ens160   Link encap:Ethernet  HWaddr 00:0c:29:bf:e1:df
         inet addr:192.168.30.128  Bcast:192.168.30.255
         Mask:255.255.255.0
```

在 Xshell 会话配置中设置 Ubuntu 操作系统的用户名和密码，单击“连接”按钮，即可连接前面安装好的 Ubuntu 操作系统，如图 2-8 所示。

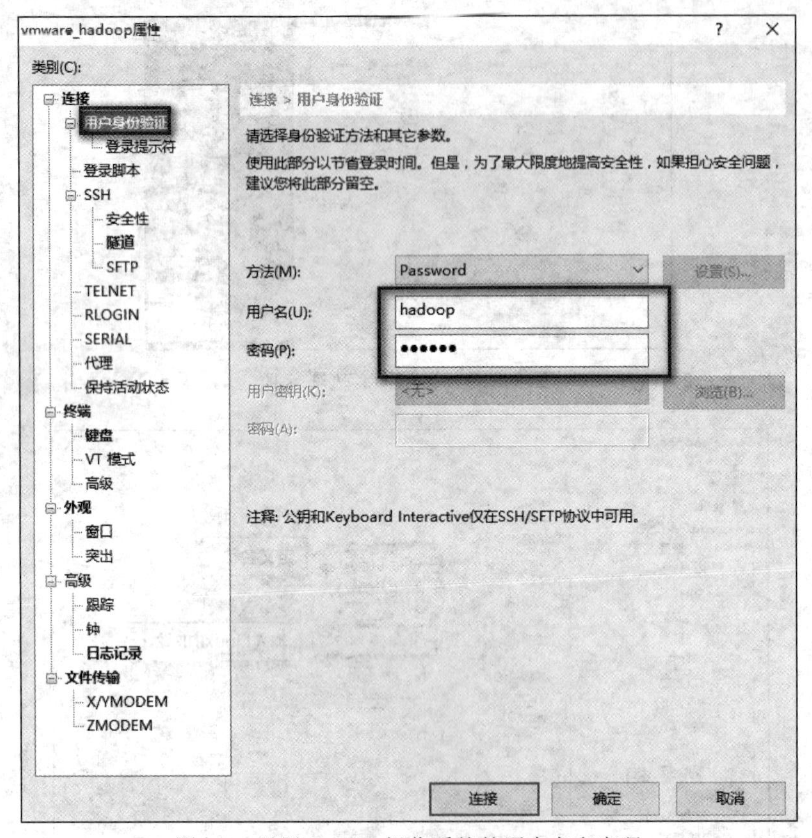

图 2-8　设置 Ubuntu 操作系统的用户名和密码

2.2.4　安装 Java

安装 Java 的具体步骤如下。

步骤 1：下载 jdk 安装包。

进入 Oracle 官网，选择 JDK8 安装包，将其下载到 Windows 操作系统的目录下，例如 D:\soft。之后通过 Xftp 将 JDK 安装包上传至 Ubuntu 系统主目录。本书使用的 JDK8 安装包为 jdk-8u381-linux-x64.tar.gz。

步骤 2：解压安装包。

解压安装包至主目录，命令如下。

```
$cd  ~
$tar  -zxvf jdk-8u381-linux-x64.tar.gz
```

步骤 3：创建软连接。

创建软连接的命令如下。

```
$ln  -s  jdk1.8.0_381 jdk1.8
```

步骤 4：配置环境变量。

首先，为了可以在任意目录下使用 Java 相关命令，我们需要告诉操作系统 Java 的命令在哪些目录下。我们通过以下命令在～/.bashrc 文件中配置 PATH 的环境变量，这样系统会在配置的目录下查找相关命令。

```
$vi  ~/.bashrc
```

其次，打开上述文件，在其末尾添加以下代码，之后保存并退出。

```
export JAVA_HOME = ~/jdk1.8
export CLASSPATH = .:$JAVA_HOME/lib/dt.jar:$JAVA_HOME/lib/tools.jar
export PATH = $PATH:$JAVA_HOME/bin
```

再次，使配置生效，命令如下。

```
$source  ~/.bashrc
```

最后，验证 Java 环境变量配置是否正确的命令如下。

```
$java -version
java version "1.8.0_381"
Java(TM) SE Runtime Environment (build 1.8.0_381-b09)
Java HotSpot(TM) 64-Bit Server VM (build 25.381-b09, mixed mode)
```

2.2.5　安装 Hadoop

步骤 1：下载 Hadoop 安装包。

进入 Apache 官网，选择对应版本的 Hadoop 安装包，将其下载到 Windows 操作系统的目录下，例如 D:\soft。之后通过 Xftp 将 Hadoop 安装包上传至 Ubuntu 操作系统主目录，本书使用的 Hadoop 安装包为 hadoop-3.2.0.tar.gz。

步骤 2：解压安装包。

解压安装包至主目录，命令如下。

```
$cd  ~
$tar  -zxvf hadoop-3.2.0.tar.gz
```

步骤 3：创建软连接，命令如下。

```
$ln  -s  hadoop-3.2.0  hadoop
```

步骤 4：配置环境变量。

首先，为了可以在任意目录下使用 Hadoop 相关命令，需要告诉操作系统 Hadoop 的命令在哪些目录下。我们通过以下命令在～/.bashrc 文件中配置 PATH 的环境变量，这样系统会在配置的目录下查找相关命令。

```
$vi  ~/.bashrc
```

打开上述文件，在其末尾添加以下代码，之后保存并退出。

```
export HADOOP_HOME  = ~/hadoop
export PATH = $PATH:$HADOOP_HOME/bin:$HADOOP_HOME/sbin
```

然后，使配置生效，命令如下。

```
$source  ~/.bashrc
```

最后，验证 Hadoop 环境变量配置是否正确的命令如下。

```
$ hadoop version
Hadoop 3.2.0
Compiled by sunilg on 2019-01-08T06:08Z
Compiled with protoc 2.5.0
```

```
From source with checksum d3f0795ed0d9dc378e2c785d3668f39
This command was run using /home/ubuntu/hadoop-3.2.0/share/hadoop/common/hadoop-common-3.2.0.jar
```

解压后的～/hadoop-3.2.0 目录的说明如下。

```
hadoop-3.2.0/
├── bin   // 存储操作命令，hdfs/hadoop 在这里
├── etc
│   └── Hadoop   ---所有配置文件
├── include
├── lib    ---本地库（native 库，一些动态库）
├── libexec
├── LICENSE.txt
├── logs   ---日志（默认是没有的，只有 Hadoop 运行后才会产生）
├── NOTICE.txt
├── README.txt
├── sbin   ---集群的命令，例如启动、停止
└── share
├── doc   ---文档
└── hadoop   ---所有依赖的 jar 包
        ├── client
        │   ├── hadoop-client-api-3.2.0.jar
        │   ├── hadoop-client-minicluster-3.2.0.jar
        │   └── hadoop-client-runtime-3.2.0.jar
        ├── common
        │   ├── hadoop-common-3.2.0.jar
        │   ├── hadoop-common-3.2.0-tests.jar
        │   ├── hadoop-kms-3.2.0.jar
        │   ├── hadoop-nfs-3.2.0.jar
        │   ├── jdiff
        │   ├── lib
        │   ├── sources
        │   └── webapps
        ├── hdfs
        │   ├── hadoop-hdfs-3.2.0.jar
        │   ├── hadoop-hdfs-3.2.0-tests.jar
        │   ├── hadoop-hdfs-client-3.2.0.jar
        │   ├── hadoop-hdfs-client-3.2.0-tests.jar
        │   ├── hadoop-hdfs-httpfs-3.2.0.jar
        │   ├── hadoop-hdfs-native-client-3.2.0.jar
        │   ├── hadoop-hdfs-native-client-3.2.0-tests.jar
        │   ├── hadoop-hdfs-nfs-3.2.0.jar
        │   ├── hadoop-hdfs-rbf-3.2.0.jar
        │   ├── hadoop-hdfs-rbf-3.2.0-tests.jar
        │   ├── jdiff
        │   ├── lib
        │   ├── sources
        │   └── webapps
        ├── mapreduce
        │   ├── hadoop-mapreduce-client-app-3.2.0.jar
```

```
│   ├── hadoop-mapreduce-client-common-3.2.0.jar
│   ├── hadoop-mapreduce-client-core-3.2.0.jar
│   ├── hadoop-mapreduce-client-hs-3.2.0.jar
│   ├── hadoop-mapreduce-client-hs-plugins-3.2.0.jar
│   ├── hadoop-mapreduce-client-jobclient-3.2.0.jar
│   ├── hadoop-mapreduce-client-jobclient-3.2.0-tests.jar
│   ├── hadoop-mapreduce-client-nativetask-3.2.0.jar
│   ├── hadoop-mapreduce-client-shuffle-3.2.0.jar
│   ├── hadoop-mapreduce-client-uploader-3.2.0.jar
│   ├── hadoop-mapreduce-examples-3.2.0.jar
│   ├── jdiff
│   ├── lib
│   ├── lib-examples
│   └── sources
├── tools
└── yarn
    ├── hadoop-yarn-api-3.2.0.jar
    ├── hadoop-yarn-applications-distributedshell-3.2.0.jar
    ├── hadoop-yarn-applications-unmanaged-am-launcher-3.2.0.jar
    ├── hadoop-yarn-client-3.2.0.jar
    ├── hadoop-yarn-common-3.2.0.jar
    ├── hadoop-yarn-registry-3.2.0.jar
    ├── hadoop-yarn-server-applicationhistoryservice-3.2.0.jar
    ├── hadoop-yarn-server-common-3.2.0.jar
    ├── hadoop-yarn-server-nodemanager-3.2.0.jar
    ├── hadoop-yarn-server-resourcemanager-3.2.0.jar
    ├── hadoop-yarn-server-router-3.2.0.jar
    ├── hadoop-yarn-server-sharedcachemanager-3.2.0.jar
    ├── hadoop-yarn-server-tests-3.2.0.jar
    ├── hadoop-yarn-server-timeline-pluginstorage-3.2.0.jar
    ├── hadoop-yarn-server-web-proxy-3.2.0.jar
    ├── hadoop-yarn-services-api-3.2.0.jar
    ├── hadoop-yarn-services-core-3.2.0.jar
    ├── hadoop-yarn-submarine-3.2.0.jar
    ├── lib
    ├── sources
    ├── test
    ├── timelineservice
    ├── webapps
    └── yarn-service-examples
```

2.2.6 克隆主机

下面演示如何搭建完全分布模式。这里需要 3 台主机，我们可以利用第 1 台主机克隆出 3 台主机。读者如果不考虑搭建完全分布模式，则可以跳过本小节内容。克隆主机的具体步骤如下。

步骤 1：单击 Ubuntu-VMware Workstation 界面上方的倒三角形图标，在下拉菜单中单击“关闭客户机”选项，如图 2-9 所示。这是因为如果不先关闭 Ubuntu 主机（客户机），就无法进行克隆操作。

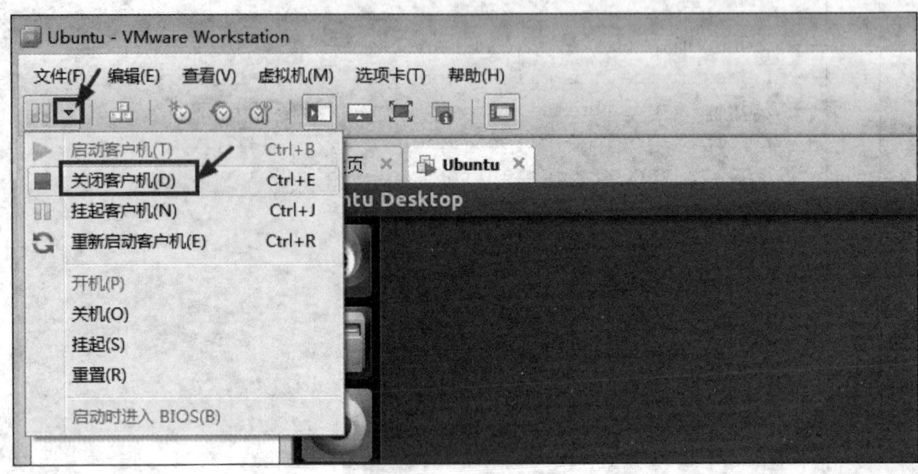

图 2-9 关闭客户机

步骤 2：在左侧栏选择要克隆的主机“Ubuntu”，并单击鼠标右键，在弹出的界面中单击“管理”→“克隆”，如图 2-10 所示。

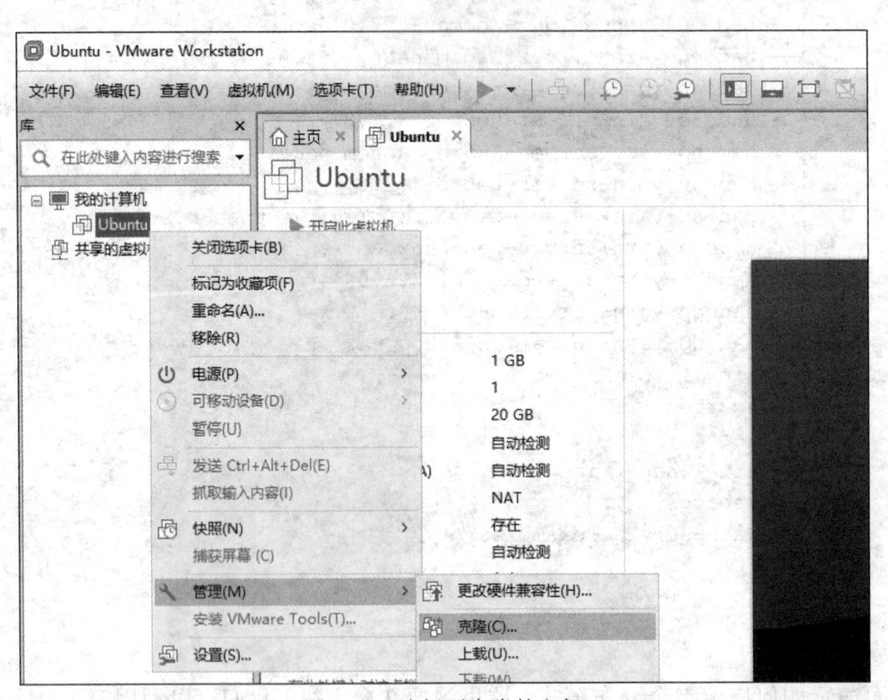

图 2-10 选择要克隆的主机

步骤 3：连续单击“下一步”按钮至“克隆类型”界面，在该界面中选择“创建完整克隆”选项，并单击“下一步”按钮，如图 2-11 所示。

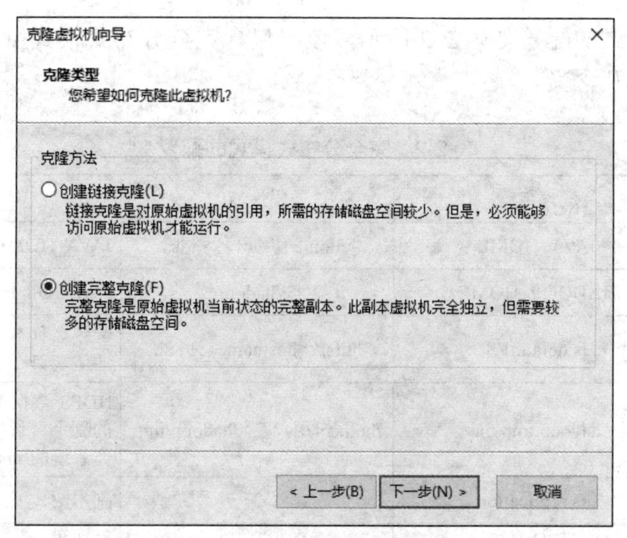

图 2-11 选择“创建完整克隆”

步骤 4：在“新虚拟机名称”界面修改虚拟机名称和位置，如图 2-12 所示。每台克隆机需要安装在一个空的文件夹中。

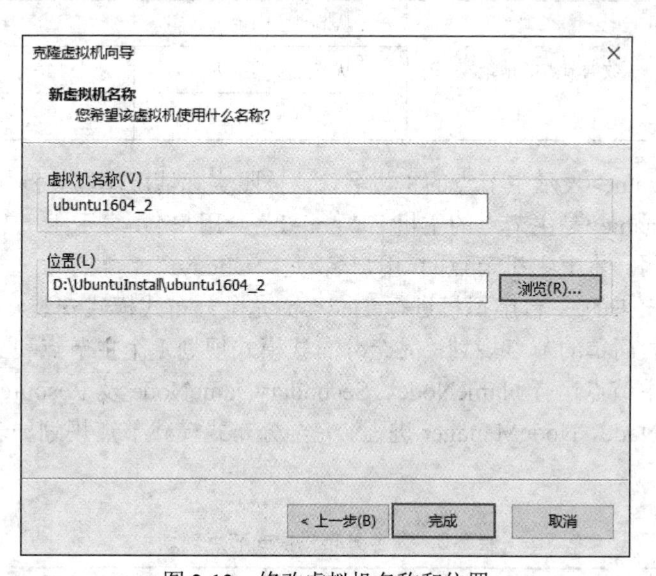

图 2-12 修改虚拟机名称和位置

步骤 5：在图 2-12 所示界面单击“完成”按钮，等待克隆完成。此过程需要 3～5 min。

步骤 6：重复执行以上步骤，直至克隆出 3 台主机为止。

2.2.7 安装完全分布式模式

完全分布式模式也叫集群模式，是真正的分布式的、由 3 个及以上的实体机或者虚拟机组成的集群。

完全分布式模式的配置如表 2-2 所示，该表包含完全分布式模式所需要配置的文件、属性名称、属性值及含义。

表 2-2　完全分布式模式的配置

文件名称	属性名称	属性值	含义
hadoop-env.sh	JAVA_HOME	/home/<用户名>/jdk	JAVA_HOME
.bashrc	HADOOP_HOME	～/hadoop	HADOOP_HOME
core-site.xml	fs.defaultFS	hdfs://<hostname>: 9820	配置名称节点地址，9820 是 RPC 通信端口
	hadoop.tmp.dir	/home/<用户名>/hadoop/tmp	HDFS 数据保存在 Linux 操作系统的哪个目录，默认目录是 Linux 操作系统的/tmp 目录
hdfs-site.xml	dfs.replication	3	副本数，默认值是 3
mapred-site.xml	mapreduce.framework.name	yarn	配置为 yarn 时表示集群模式，配置为 local 时表示本地模式
yarn-site.xml	yarn.resourcemanager.hostname	<hostname>	ResourceManager 的主机名
	yarn.nodemanager.aux-services	mapreduce_shuffle	NodeManager 上运行的附属服务
workers	数据节点的地址	从节点 1 的主机名	
		从节点 2 的主机名	
		从节点 3 的主机名	

注意：<hostname>表示主节点的主机名，请按照实际情况填写，例如，本书的完全分布式模式中<hostname>是主节点的主机名“node1”；<用户名>表示主机节点的用户名，请按照实际情况填写，本书主机节点的<用户名>是“Ubuntu”。

下面对克隆出来的 3 台虚拟机进行 Hadoop 完全分布式模式安装，并查看主机的 IP 地址。安装前先做简单的节点规划：完全分布式模式规划 1 个主节点和 2 个从节点，共 3 个节点，其中，主节点运行 NameNode、SecondaryNameNode 及 ResourceManager 进程，从节点运行 DataNode、NodeManager 进程。完全分布式模式节点规划如表 2-3 所示，具体步骤如下。

表 2-3　完全分布式模式节点规划

主机名称	IP 地址	节点角色	运行进程
node1	192.168.30.131	主节点	NameNode，SecondaryNameNode，DataNode，NodeManager，ResourceManager
node2	192.168.30.132	从节点	DataNode，NodeManager
node3	192.168.30.133	从节点	DataNode，NodeManager

步骤 1：安装前准备。

这里采用 3 台克隆后的虚拟机来安装 Hadoop 完全分布式模式，使用 ifconfig 命令查看IP地址，IP地址依次为192.168.30.131、192.168.30.132、192.168.30.133。我们建议设置IP地址为静态IP地址。

步骤2：修改主机名。

首先，修改第一台克隆主机的名称为node1。在虚拟机上打开第一台克隆主机，用vi命令编辑 /etc/hostname 文件，如下所示。

```
$sudo  vi  /etc/hostname
```

将原有内容删除，添加如下内容。

```
node1
```

重启这台虚拟机，使之生效，命令如下。

```
$sudo  reboot
```

然后，修改第二台克隆主机的名称为node2。在虚拟机上打开第二台克隆主机，用vi命令编辑/etc/hostname文件，如下所示。

```
$sudo  vi /etc/hostname
```

将原有内容删除，添加如下内容。

```
node2
```

重启这台虚拟机，使之生效，命令如下。

```
$sudo  reboot
```

最后，修改第三台克隆主机的名称为node3。在虚拟机上打开第三台克隆主机，用vi命令编辑/etc/hostname文件，如下所示。

```
$sudo  vi  /etc/hostname
```

将原有内容删除，添加如下内容。

```
node3
```

重启这台虚拟机，使之生效，命令如下。

```
$sudo  reboot
```

步骤3：映射IP地址及主机名。

依次修改这3台虚拟机的/etc/hosts文件，命令如下。

```
$sudo  vi  /etc/hosts
```

在文件末尾添加以下内容，注意，IP地址要根据实际情况填写。

```
192.168.30.131    node1
192.168.30.132    node2
192.168.30.133    node3
```

步骤4：免密登录设置。

如果只需要通过本主机登录别的主机，把本主机当作客户端，则在本主机安装SSH客户端（openssh-client）软件即可。如果要让别的主机（包括本主机自己）登录本主机，也就是说把本主机当作服务端，就需要安装SSH服务端（openssh-server）软件。Ubuntu操作系统默认没有安装SSH服务端软件，SSH服务端软件的安装请参考前文内容。

登录其他主机时通常需要输入密码。如果要让普通用户不需要输入密码就可以登录集群内的主机（即免密登录），那么需要进行如下设置。

在完全分布式模式下，集群内任一主机可以通过设置实现免密登录集群内的所有主机，即两两免密登录。这种设置并不复杂，先分别在node1、node2、node3主机上生成密钥对（公钥–私钥），然后将公钥发送给集群内的所有主机即可。下面以node1主机免密登录集群内其他所有主机为例进行讲解，其他两台主机可按照node1的步骤完成免密登录设置。

首先，在node1主机上生成密钥对，命令如下。

```
$ssh-keygen  -t  rsa
```

其中，rsa表示加密算法。键入上述命令后连续按3次回车键，系统会自动在~/.ssh目录下生成公钥（id_rsa.pub）和私钥（id_rsa）。这时，可通过命令$ls ~/.ssh 查看，具体如下。

```
$ls  ~/.ssh
id_rsa  id_rsa.pub
```

然后，将node1主机的公钥id_rsa.pub复制到node1、node2和node3主机上，命令如下。

```
$ssh-copy-id  -i  ~/.ssh/id_rsa.pub node1
$ssh-copy-id  -i  ~/.ssh/id_rsa.pub node2
$ssh-copy-id  -i  ~/.ssh/id_rsa.pub node3
```

最后，验证免密登录。在node1主机上输入以下命令进行验证。

```
$ssh  node1
$ssh  node2
$ssh  node3
```

步骤5：安装NTP服务。

完全分布式模式由多台主机组成，各台主机的时间可能存在较大差异，如果差异较大，那么在执行MapReduce程序时会出现问题。网络时间协议（Network Time Protocol，NTP）服务通过获取网络时间使集群内不同主机的时间保持一致，Ubuntu 操作系统默认没有安装NTP服务。下面在3台主机上分别安装NTP服务，命令如下（在安装NTP服务时，主机需连接互联网）。

```
$sudo  apt-get  install  ntp
```

查看NTP服务是否运行，命令如下。如果输出“ntp”，则说明NTP服务正在运行。

```
$sudo  dpkg  -l  | grep  ntp
```

步骤6：设置Hadoop配置文件。

在node1主机上进行操作，使用以下命令进入node1主机的Hadoop配置文件目录${HADOOP_HOME}/etc/hadoop。

```
$cd  ~/hadoop/etc/Hadoop
```

配置hadoop-env.sh文件。进入Hadoop配置文件所在目录，使用以下命令修改hadoop-env.sh文件。

```
$vi  hadoop-env.sh
```

在上述文件中找到export JAVA_HOME所在行，把行首的注释符号（#）删掉，之后按实际目录修改JAVA_HOME的值，命令如下。

```
export  JAVA_HOME = ~/jdk1.8              // 修改前的内容，运行时删除此行代码
export  JAVA_HOME = /home/hadoop/jdk // 修改后的内容
```

注意，JAVA_HOME = /home/hadoop/jdk 中的 hadoop 为用户名，读者按实际用户名进行修改。

配置 core-site.xml。用 vi 命令打开 core-site.xml，具体如下

```
$vi  core-site.xml
```

参考以下内容对上述文件的内容进行修改，修改完成后保存并退出。

```
<?xml version = "1.0" encoding = "UTF-8"?>
<?xml-stylesheet type = "text/xsl" href = "configuration.xsl"?>
<configuration>
    <property>
        <name>fs.defaultFS</name>
        <value>hdfs:// node1:9820</value>
        <!-- 以上 IP 地址或主机名要按实际情况修改 -->
    </property>
    <property>
        <name>hadoop.tmp.dir</name>
        <value>/home/ubuntu/hadoop/tmp</value>
    </property>
</configuration>
```

上述内容的配置说明如下。

fs.defaultFS 用于指定默认文件系统的 URI，其格式一般为 hdfs://host:port，其中，host 可以设置为 Ubuntu 操作系统的 IP 地址及主机名称中的任意一个，这里设置为主机名；port 如果不配置，则使用默认端口号 9820。

hadoop.tmp.dir 用于指定 Hadoop 的临时工作目录，其格式为/home/<用户名>/hadoop/tmp，其中，<用户名>需根据实际情况修改。注意，一定要配置 hadoop.tmp.dir，否则默认的 tmp 目录在/tmp 目录下，重启 Ubuntu 操作系统时 tmp 目录下的 dfs/name 文件夹会被删除，这会导致没有 NameNode 进程。

配置 hdfs-site.xml。修改 hdfs-site.xml 文件内容为以下内容。

```
<?xml version = "1.0" encoding = "UTF-8"?>
<?xml-stylesheet type = "text/xsl" href = "configuration.xsl"?>
<configuration>
    <property>
        <name>dfs.namenode.name.dir</name>
        <value>~/hadoop/dfs/name</value>
    </property>
    <property>
        <name>dfs.datanode.data.dir</name>
        <value>~/hadoop/dfs/data</value>
    </property>
    <property>
        <name>dfs.replication</name>
        <value>3</value>
    </property>
</configuration>
```

配置 mapred-site.xml。使用以下命令复制 mapred-site.xml.template，生成 mapred-site.xml。

```
$cp  mapred-site.xml.template  mapred-site.xml
```

用 vi 命令打开 mapred-site.xml，具体如下。

```
$vi mapred-site.xml
```

将上述文件的内容修改成以下内容，之后保存并退出。

```
<?xml version = "1.0"?>
<?xml-stylesheet type = "text/xsl" href = "configuration.xsl"?>
<configuration>
    <property>
        <name>mapreduce.framework.name</name>
        <value>yarn</value>
    </property>
    <property>
        <name>yarn.app.mapreduce.am.env</name>
        <value>HADOOP_MAPRED_HOME = $HADOOP_HOME</value>
    </property>
    <property>
        <name>mapreduce.map.env</name>
        <value>HADOOP_MAPRED_HOME = $HADOOP_HOME</value>
    </property>
    <property>
        <name>mapreduce.reduce.env</name>
        <value>HADOOP_MAPRED_HOME = $HADOOP_HOME</value>
    </property>
</configuration>
```

在上述代码中，mapreduce.framework.name 的默认值为 local，这里设置为 yarn，目的是让 MapReduce 程序运行在 YARN 框架上。

配置 yarn-site.xml。用 vi 命令打开 yarn-site.xml，具体如下。

```
$vi  yarn-site.xml
```

将上述文件的内容修改成以下内容，之后保存并退出。

```
<?xml version = "1.0"?>
<configuration>
    <property>
        <name>yarn.resourcemanager.hostname</name>
        <value>node1</value>
        <!-- 以上主机名或 IP 地址按实际情况修改 -->
    </property>
    <property>
        <name>yarn.nodemanager.aux-services</name>
        <value>mapreduce_shuffle</value>
    </property>
</configuration>
```

在上述代码中，yarn.resourcemanager.hostname 表示资源管理器的主机，可设置为 Ubuntu 操作系统的主机名或 IP 地址；yarn.nodemanager.aux-services 表示节点管理器的辅助服务器，可设置为 mapreduce_shuffle，其默认值为空。

通过以上设置，我们完成了 Hadoop 伪分布式模式的配置。其实，Hadoop 可以配置的属性还有很多，没有配置的属性可以用默认值。默认属性配置存储在 core-default.xml、hdfs-default.xml、mapred-default.xml 和 yarn-default.xml 文件中。读者可以在 Hadoop 官网上查询对应文档或通过 locate 命令来查找文件所在路径，再通过 cat 命令查看其内容。以下代码展示了一个查看默认属性相关信息的示例。

```
$locate  core-default.xml
/home/hadoop/soft/hadoop-3.2.0/share/doc/hadoop/hadoop-project-dist/hadoop-comm
on/core-default.xml
$cat /home/hadoop/soft/hadoop-3.2.0/share/doc/hadoop/hadoop-project-dist
/hadoop-common/core-default.xml
```

配置 workers 文件，其目的是指定哪些主机是从节点。使用以下命令进入配置目录 ${HADOOP_HOME}/etc/hadoop，并修改 workers 文件。

```
$cd  ~/hadoop/etc/hadoop
$vi  workers
```

将上述文件原有的内容删除，并添加以下内容。

```
node1
node2
node3
```

分发配置。使用以下命令将 node1 主机的配置文件分发至 node2 主机和 node3 主机。

```
$cd  ~/hadoop/etc/
$scp  -r hadoop  ubuntu@node2: ~/hadoop/etc/
$scp  -r hadoop  ubuntu@node3: ~/hadoop/etc/
```

步骤 7：格式化 HDFS。

在 node1 主机上进行 HDFS 格式化操作，使用的命令如下。

```
$hdfs  namenode  -format
```

步骤 8：启动 Hadoop。

启动只需在 node1 主机上进行操作。使用以下命令分别启动 HDFS 和 YARN。

```
$start-dfs.sh
$start-yarn.sh
```

也可以用以下命令启动 HDFS 和 YARN。

```
$start-all.sh
```

步骤 9：验证 Hadoop 进程。

用 jps 命令在所有主机上验证 Hadoop 进程是否成功启动，具体如下。

```
$jps
```

在 node1 主机上运行上述命令后，得到的结果中包含以下 3 个进程，这表示成功启动了 Hadoop 进程。

```
$jps
SecondaryNameNode
NameNode
ResourceManager
NodeManager
DataNode
```

在 node2 和 node3 主机上分别执行 jps 命令，得到的结果均包含以下两个进程，这表示成功启动了Hadoop进程。

```
$jps
NodeManager
DataNode
```

如果某台主机缺少某个进程，那么应该到该主机上查找相关的日志来分析原因，日志存储在${HADOOP_HOME}/logs目录下。例如，node3主机缺少DataNode进程，那么使用以下命令到node3主机的${HADOOP_HOME}/logs目录下，查看与DataNode进程相关的日志。在日志中找到含有WARN、Error、Exception等关键字的语句，通过上网搜索这些语句查找解决问题的办法。

```
$ssh  node3
$cd  ~/hadoop-3.2.0/logs
$cat  hadoop-hadoop-datanode-node3.log
```

读者也可以通过vi命令查看日志内容，具体如下。

```
$vi  hadoop-hadoop-datanode-node1.log
```

最新出现的错误都在文件末尾处。

步骤10：测试Hadoop。

我们通过一个MapReduce程序来测试Hadoop：统计HDFS中/input/data.txt文件中单词出现的次数。

在Ubuntu操作系统的主目录下，创建一个文本文件data.txt，命令如下。

```
$cd  ~
$vi  data.txt
```

在data.txt文件中输入如下内容，之后保存并退出。

```
Hello World
Hello Hadoop
```

在HDFS中创建input文件夹，命令如下。

```
$hdfs  dfs  -mkdir  /input
```

将data.txt文件上传到HDFS的input文件夹中，命令如下。

```
$hdfs  dfs  -put  data.txt  /input
```

查看是否上传成功，命令如下。

```
$hdfs  dfs  -ls  /input
Found 1 items
-rw-r--r--  1  hadoop supergroup   25 2018-10-13 22:40  /input/data.txt
```

运行MapReduce WordCount例子，命令如下。

```
$cd  ~/hadoop/share/hadoop/mapreduce
$hadoop  jar  hadoop-mapreduce-examples-3.2.0.jar  wordcount /input/data.txt /output
```

说明：第二条命令和第一条命令在同一行。

查看结果，命令和得到的结果如下。

```
$hdfs  dfs  -cat  /output/part-r-00000
Hadoop    1
Hello     2
World     1
```

步骤 11：停止 Hadoop 进程。

如果要关闭 Hadoop 进程，可采用下列命令分别关闭 HDFS 和 YARN。

```
$stop-dfs.sh
$stop-yarn.sh
```

读者也可以使用以下命令停止所有进程。

```
$stop-all.sh
```

用 jps 命令查看是否关闭了 Hadoop 所有进程，具体如下。

```
$jps
Jps
```

至此，Hadoop 完全分布式模式搭建完成。

2.3 查看 Hadoop 集群的基本信息

【任务描述】Hadoop 启动后，查询 Hadoop 集群的基本信息，其中包括存储系统信息和计算资源信息。

2.3.1 查询存储系统信息

在浏览器中输入内部网址 http://node1:9860，查看 NameNode 和 DataNode 信息，如图 2-13 所示。

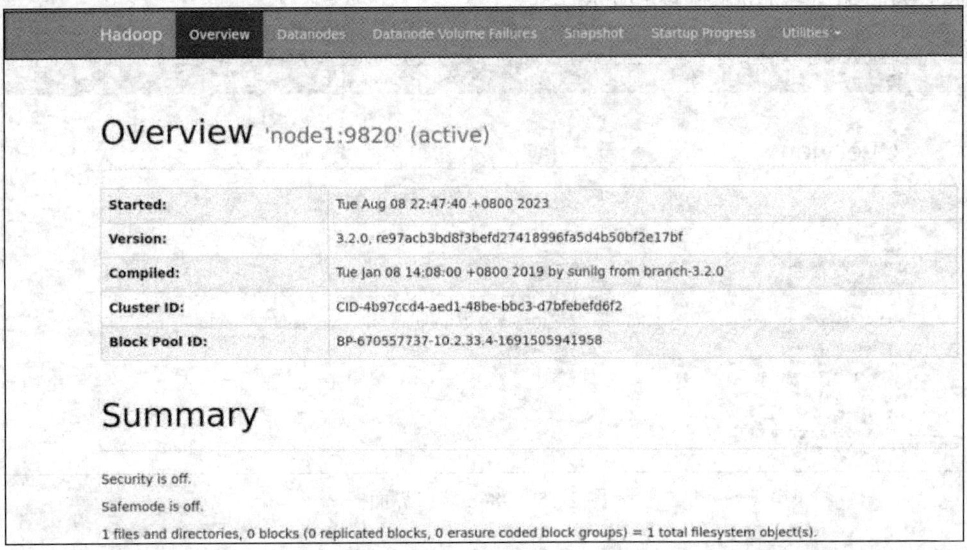

图 2-13　9860 端口界面

单击图 2-13 所示界面的“Datanodes”选项，查看 DataNode 信息。得到的结果如图 2-14 所示，可以看出有 3 个 DataNode 进程正在运行。

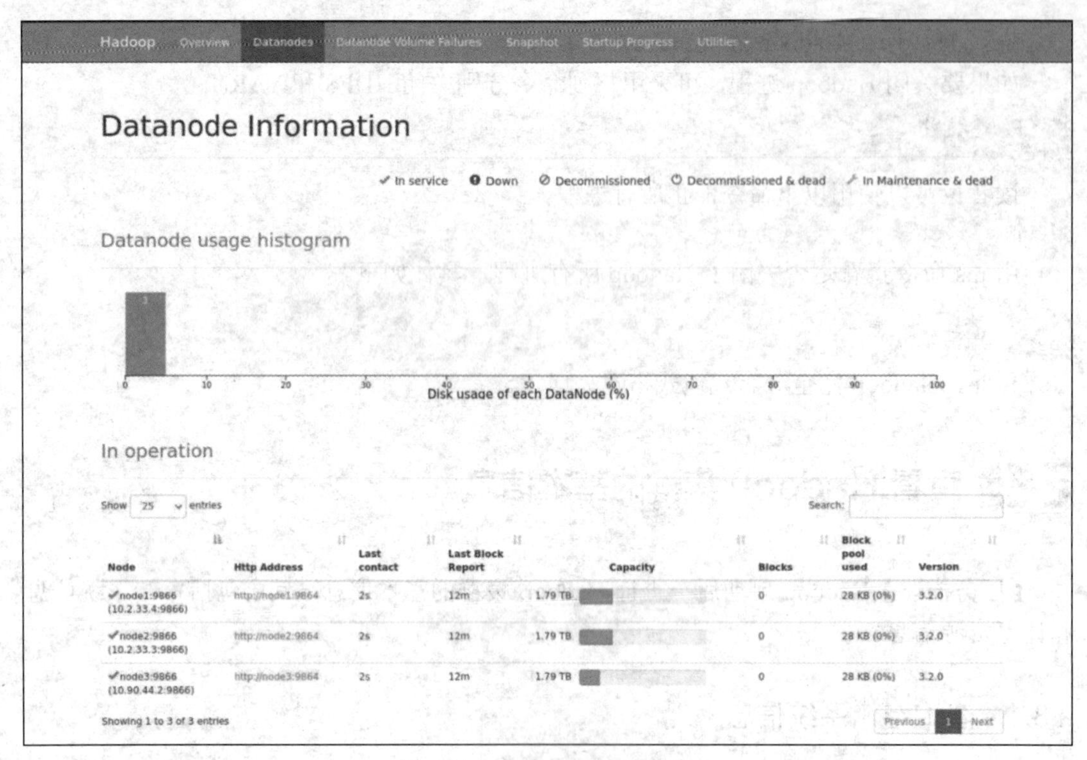

图 2-14　查看 DataNode 信息

在浏览器中输入内部网址 http://node1:9868，查看 SecondaryNameNode 信息。node1 主机的 9868 端口界面如图 2-15 所示。

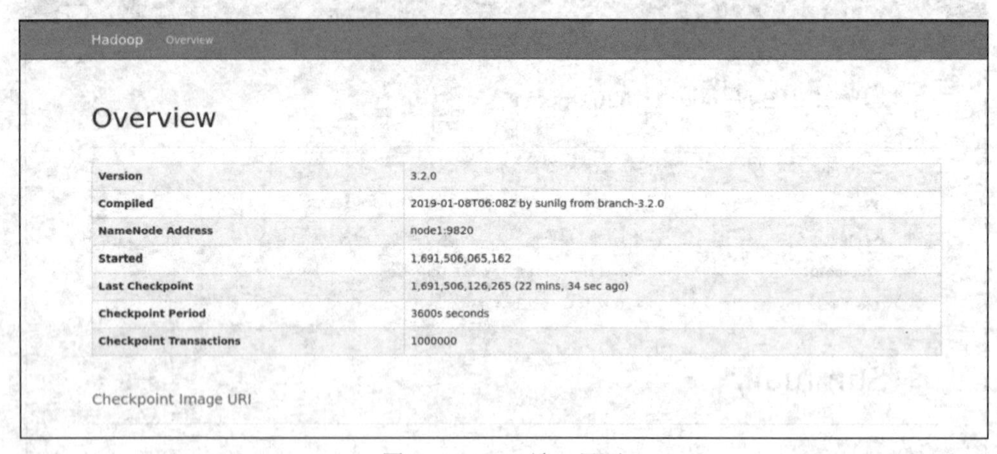

图 2-15　9868 端口界面

2.3.2　查询计算资源信息

在浏览器中输入http://node1:8088，查看集群内所有应用程序的信息，得到的结果如图2-16

所示。可以看到，Active Nodes 值为 3，这说明集群中有 3 个 NodeManager 节点正在运行。

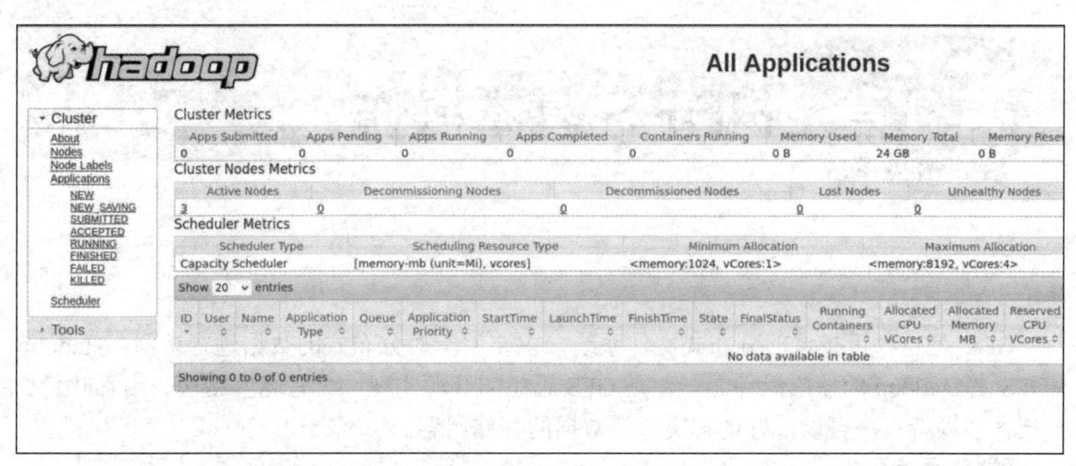

图 2-16　8088 端口界面

2.4 本章小结

本章详细讲述了 Hadoop 完全分布式集群的搭建，其中包括 Hadoop 软件的安装和配置，以及平台运行所需的环境设置和相关配套软件。同时，本章还介绍了通过浏览器查看 Hadoop 集群基本信息的方法，为后续内容提供了运行环境的搭建方法。

第3章 HDFS基本操作

GFS 是一种可扩展的分布式文件系统，主要用于对分布式的数据进行访问。Hadoop HDFS 是 Hadoop 的核心组件之一，是 GFS 思想的开源实现，具有高容错性、高吞吐量、扩展性强等特点，可以很好地解决海量数据的存储问题。

【学习目标】

1．了解 HDFS 的组成。

2．熟悉 HDFS 读/写流程。

3．掌握 HDFS 操作常用 Shell 命令的用法。

4．熟练使用 Web 界面管理 HDFS。掌握 Java 访问 HDFS 的程序编写与运行方法。

3.1 Hadoop Shell 命令操作 HDFS

【任务描述】将本地文件 test.log 上传至 HDFS 目录/user/hadoop/input 下。

要完成以上任务，必须先对 HDFS 的基础知识、组成架构、读/写流程有一个全面的认识，了解 HDFS 和相关的文件系统之间的关系，并熟练地掌握对 HDFS 文件的基本操作。

3.1.1 HDFS 简介

随着数据量越来越大，一个操作系统管理的磁盘将无法存储所有的数据，那么这些数据需要被分配到更多的操作系统管理的磁盘上。但是，这种方式不方便管理和维护数据，因此，人们迫切需要一种系统来管理多台机器上的文件。分布式文件管理系统应运而生，HDFS 是分布式文件管理系统中的一种。

HDFS 用于存储文件，通过目录树来定位文件。它是一种分布式系统，由多台服务器联合起来实现数据管理和维护功能，其中的服务器有各自的角色，负责不同的工作。HDFS 适合一次写入多次读出的应用场景。在这种场景中，一个文件经过创建、写入和保存之后不需要再进行修改。

HDFS 的优点如下。

① 高容错性，数据会自动保存多个副本。HDFS 通过增加副本的形式来提高系统的

容错率，在某个副本丢失后可以自动恢复数据。

② 适合处理大数据。在数据规模上，HDFS 能够处理规模拍字节（PB）级别的数据；在文件规模上，HDFS 能够处理百万规模的文件量。

③ 可构建在低成本的机器上。HDFS 通过多副本机制来保障系统的可靠性。

HDFS 的缺点如下。

① 不适合低时延数据访问。例如对于毫秒级的存储数据，HDFS 是做不到的。

② 无法高效存储大量的小文件。如果存储大量的小文件，HDFS 会占用名称节点大量的内存来存储文件目录和块信息，这种方式是不可取的，因为名称节点的内存总是有限的。不仅如此，小文件存储的寻址时间会超过读取时间，这违背了 HDFS 的设计初衷。

③ 不支持并发写入操作和文件的随机修改。在 HDFS 中，一个文件只能由一个线程写入数据，不允许多个线程同时写入数据。HDFS 仅支持数据追加操作，不支持文件的随机修改操作。

1. HDFS 的架构

HDFS 由名称节点、数据节点、第二名称节点组成，HDFS 的基本架构是主从结构，其中，名称节点是主节点，数据节点是从节点。HDFS 架构如图 3-1 所示。

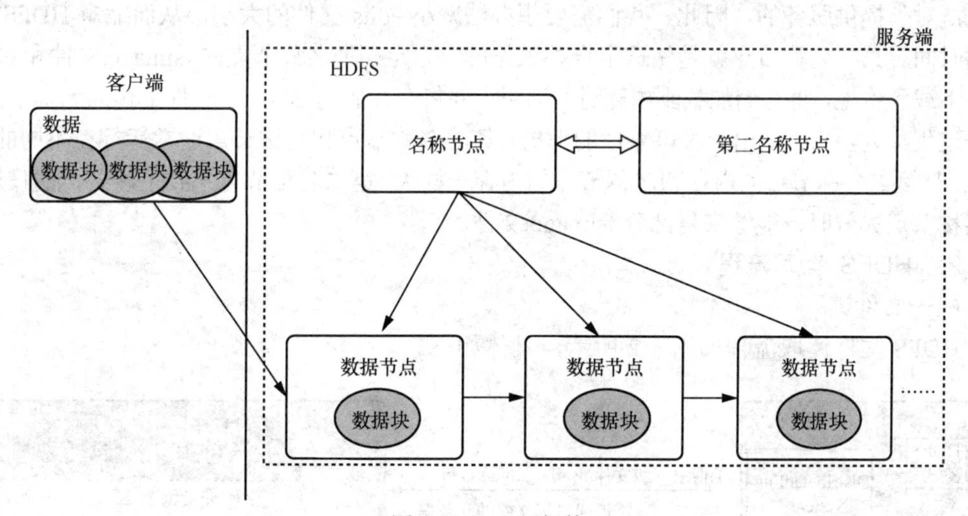

图 3-1　HDFS 架构

HDFS 先把客户端上传的大数据文件切分成若干个小的数据块，再把这些数据块以多副本（默认 3 个副本）的形式分别写入不同的节点，这些负责保存数据块的节点称为数据节点。当用户访问数据文件时，为了保证能够读取到每一个数据块，HDFS 使用一个专门保存文件属性信息的节点——名称节点。另外，HDFS 中还有另一种节点——第二名称节点，我们在后面介绍它的内容。这 3 种节点的功能如下。

名称节点是 HDFS 的管理者，它的职责有以下四方面。

① 管理 HDFS 的名称空间。

② 配置副本策略。

③ 管理数据块映射信息。

④ 处理客户端的读/写请求。

数据节点是执行命令的节点。在 HDFS 中，名称节点下达命令，数据节点执行具体操作。数据节点的职责有以下两方面。

① 存储数据块。

② 执行数据块的读/写操作。

第二名称节点并非名称节点的热备节点。当名称节点故障的时候，第二名称节点并不能马上替换名称节点并提供服务，它的职责有以下两方面。

① 辅助名称节点，分担该节点的工作，例如定期合并 fsimage 文件和 edits 文件，并推送给名称节点。第二名称节点定期把 edits 文件中最新的状态信息合并到 fsimage 文件中，从而达到缩短 HDFS 启动时间的目的。

② 在紧急情况下，可辅助恢复名称节点。

为什么要定期合并 fsimage 文件和 edits 文件呢？这里就涉及 HDFS 启动时的操作了。在启动 HDFS 时，系统会从 fsimage 文件中读取当前 HDFS 的元数据信息，同时需要将 edits 文件的状态信息合并到 fsimage 文件中，这个过程会占用很大资源。如果 edits 文件过大，那么这个合并过程会非常长，也会延长名称节点的启动时间。而在这个启动期间，HDFS 集群是无法对外提供服务的，因此，我们需要想办法减小 edits 文件的大小，从而缩短 HDFS 的启动时间。第二名称节点就是做这个合并工作的，它定期把名称节点的 fsimage 文件和 edits 文件下载到本地，将它们加载到内存进行合并，并将合并后的 fsimage 文件上传回名称节点，这个过程称为检查点。出于对可靠性的考虑，第二名称节点和名称节点通常运行在不同的机器上，且第二名称节点的内存和名称节点的内存一样大。第二名称节点起到了冷备份的作用，在名称节点失效时，能够恢复部分 fsimage 文件。

2. HDFS 读/写流程

（1）文件读取

HDFS 文件读取流程如图 3-2 所示，具体如下。

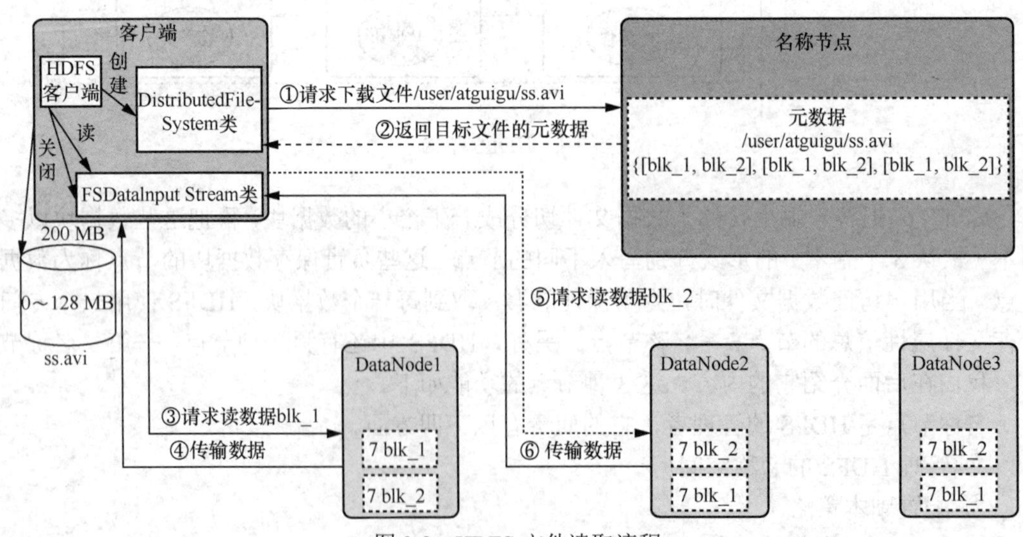

图 3-2　HDFS 文件读取流程

① 客户端通过 DistributedFileSystem 类向名称节点请求下载文件/user/atguigu/ss.avi，名称节点通过查询元数据找到数据块所在的数据节点地址。

② （按照就近原则，随机）挑选一个数据节点，请求读取数据。

③ 数据节点开始传输数据给客户端（从磁盘中读取数据输入流，以数包据（Packet）为单位来做校验）。

④ 客户端以数据包为单位接收数据，并将数据先缓存在本地，然后写入目标文件。

读取文件时，客户端需要调用 FSDataInputStream 类对数据进行读取。FSDataInputStream 类的常用方法如表 3-1 所示。

表 3-1　FSDataInputStream 类的常用方法

方法名	返回值类型	说明
read(ByteBuffer buf)	整型（int）	读取数据并扩大 buf 缓冲区，返回所读取的字节数
read(long pos, byte[] buf, int offset, int len)	整型（int）	从输入流的指定位置开始，把数据读入缓冲区。pos 参数指定从输入流中读取数据的位置，offset 参数指定数据写入缓冲区的位置（偏移量），len 参数指定读操作的最大字节数
readFully(long pos, byte[] buf)	空（void）	从指定位置开始，将所有数据读取到缓冲区
seek(long offset)	空（void）	指向输入流的第 offset 个字节
releaseBuffer(ByteBuffer buf)	空（void）	删除指定的缓冲区

【例 3-1】 编写程序，读取 HDFS 上/mydir/data.txt 文件中的内容。

```
// 读文件
import java.net.URI;
import org.apache.hadoop.conf.Configuration;
import org.apache.hadoop.fs.FSDataInputStream;
import org.apache.hadoop.fs.FileSystem;
import org.apache.hadoop.fs.Path;
public class App{
    public static void main(String[] args) throws Exception {
        Configuration conf = new Configuration();
        // 配置连接信息，名称节点地址
        URI uri = new URI("hdfs:// 192.168.30.121:8020");
        // 指定用户名，获取 FileSystem 对象
        FileSystem fs = FileSystem.get(uri,conf,"hadoop");
        Path src = new Path("/mydir/data.txt");
        FSDataInputStream dis = fs.open(src);
        String str = null;
        while( (str = dis.readLine()) != null){
        System.out.println(str);
        }
        dis.close();
```

```
        // 当不需要操作 FileSystem 时，关闭 FileSystem 对象
        fs.close();
    }
}
```

程序运行后的输出结果如下。

```
Hello HDFS !
```

（2）文件写入

HDFS 文件写入流程如图 3-3 所示，具体如下。

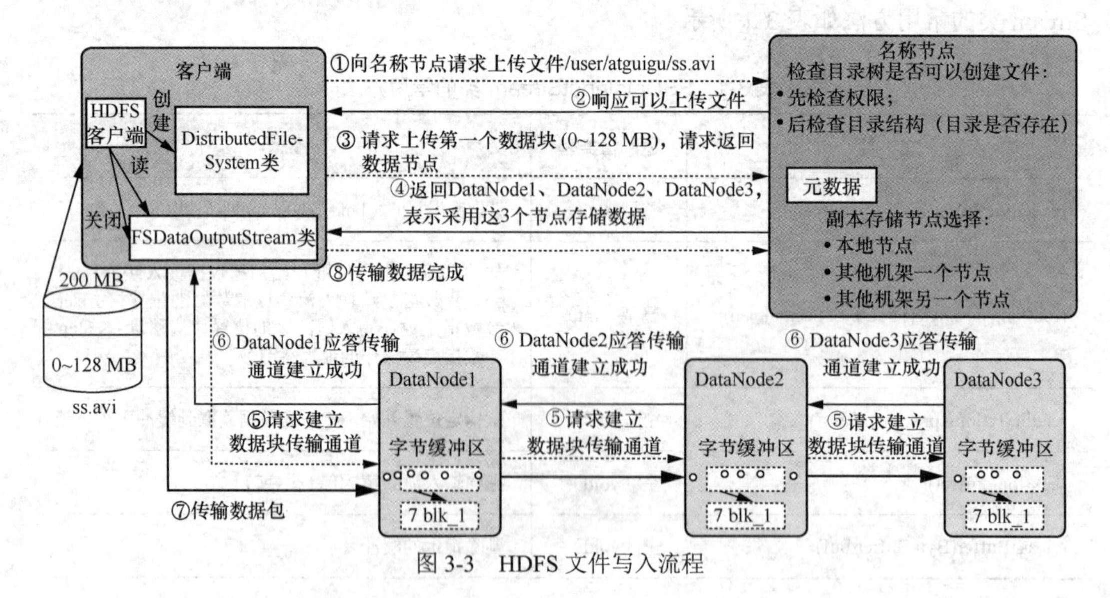

图 3-3　HDFS 文件写入流程

① 客户端通过 DistributedFileSystem 类向名称节点请求上传文件/user/atguigu/ss.avi。

② 名称节点检查目标文件是否已存在，父目录是否存在后，响应可以上传文件。

③ 客户端请求上传第一个数据块，并请求返回数据节点。

④ 名称节点返回 3 个数据节点，它们分别是 DataNode1、DataNode2、DataNode3。

⑤ 客户端通过 FSDataOutputStream 类请求 DataNode1 建立传输通道。DataNode1 收到请求继续调用 DataNode2，请求建立传输通道。之后 DataNode2 调用 DataNode3，将这个传输管道建立完成。

⑥ DataNode1、DataNode2、DataNode3 逐级应答客户端传输通道建立成功。

⑦ 客户端以数据包为单位开始向 DataNode1 上传第一个数据块（先从磁盘读取数据放到一个字节缓冲区缓存），DataNode1 收到一个数据包会传给 DataNode2，DataNode2 再传给 DataNode3。DataNode1 每传送一个数据包就会放入一个应答队列等待应答。

⑧ 当一个数据块传输完成后，客户端告诉名称节点传输数据完成并请求名称节点上传第二个数据块，并重复执行步骤③～⑦。

写入文件时，客户端需要调用 FSDataOutputStream 类对数据进行写入。FSDataOutputStream 类的常用方法如表 3-2 所示。

表 3-2　FSDataOutputStream 类的常用方法

方法名	返回值类型	说明
write(byte[] b)	空（void）	将数组 b 中的所有字节写入输出流
write(byte[] buf, int off, int len)	空（void）	将字节数组写入底层输出流，写入的字节从 off 偏移量开始，写入长度为 len
flush()	空（void）	刷新数据输出流。缓冲区内容被强制写入数据流

【例 3-2】编写程序，在 HDFS 的/mydir/data.txt 文件中写入“Hello HDFS！”。

```
import java.net.URI;
import org.apache.hadoop.conf.Configuration;
import org.apache.hadoop.fs.FSDataOutputStream;
import org.apache.hadoop.fs.FileSystem;
import org.apache.hadoop.fs.Path;
public class App{
    public static void main(String[] args) throws Exception {
        Configuration conf = new Configuration();
        // 配置连接信息，名称节点地址
        URI uri = new URI("hdfs:// 192.168.30.121:8020");
        // 指定用户名，获取 FileSystem 对象
        FileSystem fs = FileSystem.get(uri,conf,"hadoop");
        Path path = new Path("/mydir/data.txt");
        byte[] buff = "Hello HDFS！".getBytes();
        FSDataOutputStream dos = fs.create(path);
        dos.write(buff);
        // 关闭数据流
        dos.close();
        // 不需要再操作 FileSystem 对象，关闭 FileSystem 对象
        fs.close();
    }
}
```

我们通过以下 HDFS 命令查看结果。

```
$hdfs dfs -cat /mydir/data.txt
```

得到的结果如下。

```
Hello HDFS!
```

从以上例子可以看出，从 HDFS 中读取文件或者向 HDFS 文件写入数据的过程都是通过数据流完成的。HDFS 提供了数据流的输入/输出操作类，例如 FSDataInputStream 和 FSDataOutputStream。

3.1.2　HDFS Shell 命令简介

HDFS Shell 命令由一系列类似 Linux Shell 的命令组成，主要用于对 HDFS 进行目录与文件的查看、创建、删除、移动、复制、上传、下载等操作。Hadoop 支持很多 Shell

命令，例如 hadoop fs、hadoop dfs 和 hdfs dfs，这 3 种命令既有联系也有区别。

① hadoop fs：适用于任何文件系统，例如本地文件系统和 HDFS。

② hadoop dfs：只适用于 HDFS。

③ hdfs dfs：与 hadoop dfs 命令的作用一样，也只适用于 HDFS。

本书统一使用 hdfs dfs 命令对 HDFS 进行操作。

HDFS Shell 命令大致可分为操作命令、管理命令以及其他命令。文件的创建、删除、复制、移动、查找等操作需使用操作命令，操作命令以 hdfs dfs 开头。我们在终端输入 hdfs dfs 命令，查看系统支持哪些操作命令，如图 3-4 所示。

```
ubuntu@8ed18e7d903b:~$ hdfs dfs
Usage: hadoop fs [generic options]
        [-appendToFile <localsrc> ... <dst>]
        [-cat [-ignoreCrc] <src> ...]
        [-checksum <src> ...]
        [-chgrp [-R] GROUP PATH...]
        [-chmod [-R] <MODE[,MODE]... | OCTALMODE> PATH...]
        [-chown [-R] [OWNER][:[GROUP]] PATH...]
        [-copyFromLocal [-f] [-p] [-l] <localsrc> ... <dst>]
        [-copyToLocal [-p] [-ignoreCrc] [-crc] <src> ... <localdst>
        [-count [-q] [-h] <path> ...]
        [-cp [-f] [-p | -p[topax]] <src> ... <dst>]
        [-createSnapshot <snapshotDir> [<snapshotName>]]
        [-deleteSnapshot <snapshotDir> <snapshotName>]
        [-df [-h] [<path> ...]]
        [-du [-s] [-h] <path> ...]
        [-expunge]
        [-find <path> ... <expression> ...]
        [-get [-p] [-ignoreCrc] [-crc] <src> ... <localdst>]
        [-getfacl [-R] <path>]
        [-getfattr [-R] {-n name | -d} [-e en] <path>]
        [-getmerge [-nl] <src> <localdst>]
        [-help [cmd ...]]
        [-ls [-d] [-h] [-R] [<path> ...]]
        [-mkdir [-p] <path> ...]
        [-moveFromLocal <localsrc> ... <dst>]
```

图 3-4　系统支持的操作命令

从图 3-4 中可以看出，操作命令的格式为：hdfs dfs [generic options]，其中，hdfs 是 Hadoop HDFS 在 Linux 操作系统中的主命令；dfs 是子命令，表示执行文件系统操作；generic options（通用选项）由 HDFS 操作命令和操作参数组成。我们可以使用 help 命令查看给定命令的使用方法，例如使用 hdfs dfs -help put 命令查看 put 命令的具体用法，如图 3-5 所示。

```
ubuntu@8ed18e7d903b:~$ hdfs dfs -help put
-put [-f] [-p] [-l] <localsrc> ... <dst> :
  Copy files from the local file system into fs. Copying fails
 if the file already
  exists, unless the -f flag is given.
  Flags:


  -p  Preserves access and modification times, ownership and t
he mode.
  -f  Overwrites the destination if it already exists.

  -l  Allow DataNode to lazily persist the file to disk. Force
s
        replication factor of 1. This flag will result in red
uced
        durability. Use with care.
ubuntu@8ed18e7d903b:~$
```

图 3-5　查看 put 命令的具体用法

3.1.3　目录操作

对 HDFS 进行的操作主要是目录操作和文件操作，我们先介绍常用的目录操作命令。常用目录的操作主要有创建、查看和删除目录，具体命令如表 3-3 所示。

表 3-3　常用的目录操作命令

命令	说明
hdfs dfs -mkdir [-p] <paths>	创建文件夹。 [-p]：表示如果父目录不存在，则先创建父目录
hdfs dfs -ls [-d][-h][-R] <paths>	列出指定的文件和目录。 [-d]：返回 path。 [-h]：表示按照人性化的单位显示文件大小，默认单位为字节（B），其中的 h 指“human-readable”。 [-R]：级联显示 paths 下的文件
hdfs dfs -rm [-f] [-r] <src> 或者 hdfs dfs -rmdir <src>	rm 命令可以删除目录也可以删除文件，而 rmdir 命令只能删除目录。 [-f]：如果要删除的文件不存在，则不显示错误信息。 [-r]：级联删除目录下所有的文件和子目录文件

【例 3-3】创建文件夹/mydir、/mydir/dir1/dir2/dir3。

```
$ hdfs dfs -mkdir /mydir
$ hdfs dfs -mkdir -p /mydir/dir1/dir2/dir3
```

【例 3-4】列出“/”根目录下的所有文件和目录信息。

```
$ hdfs dfs -ls /
```

结果如下。

```
Found 1 items
drwxr-xr-x   - ubuntu supergroup   0 2023-08-03 04:00 /mydir/dir1
```

【例 3-5】循环列出目录、子目录及文件信息。

```
$ hdfs dfs -ls -R /mydir
```

结果如下。

```
drwxr-xr-x   - ubuntu supergroup   0 2023-08-03 04:00 /mydir/dir1
drwxr-xr-x   - ubuntu supergroup   0 2023-08-03 04:00 /mydir/dir1/dir2
drwxr-xr-x   - ubuntu supergroup   0 2023-08-03 04:00 /mydir/dir1/dir2/dir3
```

【例 3-6】删除 HDFS 中的/mydir/dir1/dir2 目录。

```
hdfs dfs -rmdir /mydir/dir1/dir2
```

3.1.4　文件操作

实际的应用中经常需要从本地文件系统向 HDFS 上传文件，或者把 DHFS 中的文件下载到本地文件系统中，有时还会涉及文件的创建和查看等操作。接下来我们对 HDFS

的文件操作命令进行详细介绍。

（1）新建文件

新建文件的命令如下。

```
hdfs dfs -touchz <paths>
```

【例 3-7】在目录/mydir 下创建大小为 0 KB 的空文件 file.txt。

```
$ hdfs dfs -touchz /mydir/file.txt
```

运行后使用 hdfs dfs -ls 命令进行查看，得到的结果如下。

```
Found 2 items
drwxr-xr-x   - ubuntu supergroup  0 2023-08-03 04:00 /mydir/dir1
-rw-r--r--   3 ubuntu supergroup  0 2023-08-03 04:23 /mydir/file.txt
```

（2）上传文件

上传文件的命令如下。

```
hdfs dfs -put [-f][-p] <localsrc> <dst>
hdfs dfs -copyFromLocal [-f][-p][-l] <localsrc> <dst>
```

[-f]：如果文件已存在，则覆盖该文件。

[-p]：递归拷贝。

put、copyFromLocal：将本地文件上传到 HDFS。

<localsrc>：表示本地文件的路径。

<dst>：表示保存在 HDFS 上的路径。

【例 3-8】在本地创建两个文件 data1.txt、data2.txt，并使用 vi 命令来编辑这两个文件，输入内容后保存并退出。之后进行如下操作。

首先，将 data1.txt 上传到 HDFS 的/mydir 目录下，采用的命令如下。

```
$ hdfs dfs -put data1.txt /mydir/data1.txt
```

或采用如下命令实现 data1.txt 的上传。

```
$ hdfs dfs -put data1.txt /mydir
```

然后，将 data2.txt 上传到 HDFS 的/mydir 目录下，采用的命令如下。

```
$ hdfs dfs -copyFromLocal data2.txt /mydir/data2.txt
```

或采用如下命令实现 data2.txt 的上传。

```
$ hdfs dfs -copyFromLocal data2.txt /mydir
```

最后，查看 HDFS 目录下文件是否上传成功，采用的命令如下。

```
$ hdfs dfs -ls
```

结果如下。

```
Found 4 items
-rw-r--r-- 3 ubuntu supergroup  26 2023-08-03 04:45 /mydir/data1.txt
-rw-r--r-- 3 ubuntu supergroup  44 2023-08-03 04:45 /mydir/data2.txt
drwxr-xr-x - ubuntu supergroup  0 2023-08-03 04:00  /mydir/dir1
-rw-r--r-- 3 ubuntu supergroup  0 2023-08-03 04:23  /mydir/file.txt
```

（3）下载文件

下载文件的命令如下。

```
hdfs dfs -get [-p] <src> <localdst>
hdfs dfs -copyToLocal [-p][-ignoreCrc][-crc] <src> <localdst>
```

get、copyToLocal：把文件从 HDFS 复制到本地。

<src>：表示 HDFS 中文件的完整路径。

<localdst>：表示要保存在本地的文件名或文件夹的路径。

【例 3-9】将 HDFS 中的/mydir/data1.txt 文件下载并保存为本地的~/local_data1.txt。

```
$ hdfs dfs -get /mydir/data1.txt ~/local_data1.txt
```

也可以采用以下命令实现上述操作。

```
$ hdfs dfs -copyToLocal /mydir/data1.txt ~/local_data1.txt
```

（4）查看文件内容

查看文件内容的命令如下。

```
hdfs dfs -text[-ignoreCrc] <src>
hdfs dfs -cat[-ignoreCrc] <src>
hdfs dfs -tail [-f] <file>
```

[-ignoreCrc]：忽略循环检验失败的文件。

[-f]：动态更新显示数据，如查看某个不断增加内容的文件的日志文件。

以上 3 个命令都可以在命令行窗口查看指定文件内容，区别在于 text 命令不仅可以查看文本文件，还可以查看压缩文件和 Avro 序列化的文件，而 cat 和 tail 这两个命令不可以；tail 命令查看的是最后 1 KB 的文件（Linux 操作系统上的 tail 命令默认查看最后 10 行记录）。

【例 3-10】查看/mydir/data1.txt 文件内容。

```
$ hdfs dfs -cat /mydir/data1.txt
```

也可以采用以下命令完成上述操作。

```
$ hdfs dfs -text /mydir/data1.txt
```

结果如下。

```
Hello HDFS!
Hello Hadoop!
Hello Hadoop and HDFS!
```

（5）移动文件

移动文件的命令如下。

```
hdfs dfs -mv <src> <dst>
```

【例 3-11】将 HDFS 中的/mydir/data1.txt 文件移动到/mydir/dir1 目录下。

```
$ hdfs dfs -mv /mydir/data1.txt /mydir/dir1/data1.txt
```

结果如下。

```
-rw-r--r--  3 ubuntu supergroup  44 2023-08-03 04:45 /mydir/data2.txt
drwxr-xr-x  - ubuntu supergroup  0 2023-08-03 04:57  /mydir/dir1
-rw-r--r--  3 ubuntu supergroup  26 2023-08-03 04:45 /mydir/dir1/data1.txt
drwxr-xr-x  - ubuntu supergroup  0 2023-08-03 04:00  /mydir/dir1/dir2
drwxr-xr-x  - ubuntu supergroup  0 2023-08-03 04:00  /mydir/dir1/dir2/dir3
-rw-r--r--  3 ubuntu supergroup  0 2023-08-03 04:23  /mydir/file.txt
```

（6）复制文件

复制文件的命令如下。

```
hdfs dfs -cp [-f][-p] <src> <dst>
```

[-f]：如果目标文件存在，则强行覆盖该文件。

[-p]：保存文件的属性。

【例 3-12】将 HDFS 中的/mydir/data2.txt 文件复制为/mydir/data2_copy.txt 文件，命令如下。

```
$ hdfs dfs -cp /mydir/data2.txt /mydir/data2_copy.txt
```

结果如下。

```
Found 4 items
-rw-r--r--   3 ubuntu supergroup 44 2023-08-03 05:03  /mydir/data2_copy.txt
-rw-r--r--   3 ubuntu supergroup 44 2023-08-03 04:45  /mydir/data2.txt
drwxr-xr-x   - ubuntu supergroup 0  2023-08-03 04:57  /mydir/dir1
-rw-r--r--   3 ubuntu supergroup 0  2023-08-03 04:23  /mydir/file.txt
```

（7）删除文件

删除文件的命令如下。

```
hdfs dfs -rm [-f] [-r|-R] [-skipTrash] <src>
```

[-f]：如果要删除的文件不存在，则不显示提示和错误信息。

[-r|-R]：级联删除目录下所有的文件和子目录文件。

[-skipTrash]：直接删除文件，不进入回收站。

【例 3-13】删除 HDFS 中的/mydir/data2.txt 文件。

```
$hdfs dfs -rm /mydir/data2.txt
```

结果如下。

```
Found 3 items
-rw-r--r--   3 ubuntu supergroup  44 2023-08-03 05:03 /mydir/data2_copy.txt
drwxr-xr-x   - ubuntu supergroup  0 2023-08-03 04:57  /mydir/dir1
-rw-r--r--   3 ubuntu supergroup  0 2023-08-03 04:23  /mydir/file.txt
```

3.1.5 利用 Web 界面管理 HDFS

HDFS 提供了管理 HDFS 的 Web 界面。通过 Web 界面，我们可以方便地查看 HDFS 相关信息。在 Linux 操作系统自带的浏览器中输入 http://localhost:9870（内部网址）后按回车键，通过 9870 端口打开 HDFS 的 Web 界面。HDFS 的 Web 管理界面中包含了 Overview、Datanodes、Datanode Volume Failures、Snapshot、Startup Progress、Utilities 等选项，如图 3-6 所示。每个菜单选项都包含相应的管理界面，我们在这些管理界面中可以查询相应的详细信息。单击选项“Utilities”→“Browse the file system”来查看文件系统的菜单，得到图 3-7 所示界面。

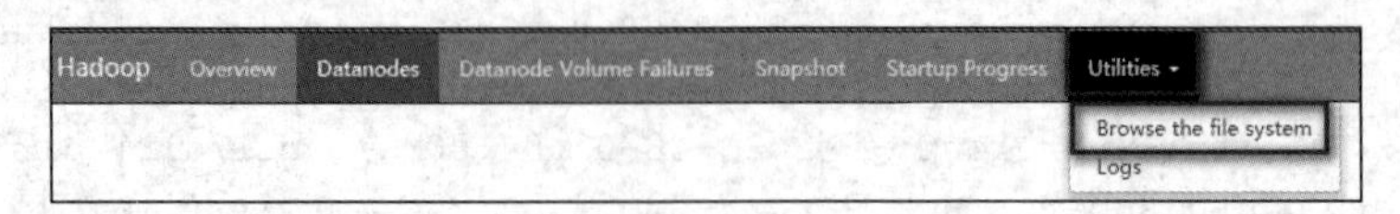

图 3-6　查看文件系统的菜单

在图 3-7 所示界面中，我们可以查看 HDFS 的目录结构及文件属性等信息。

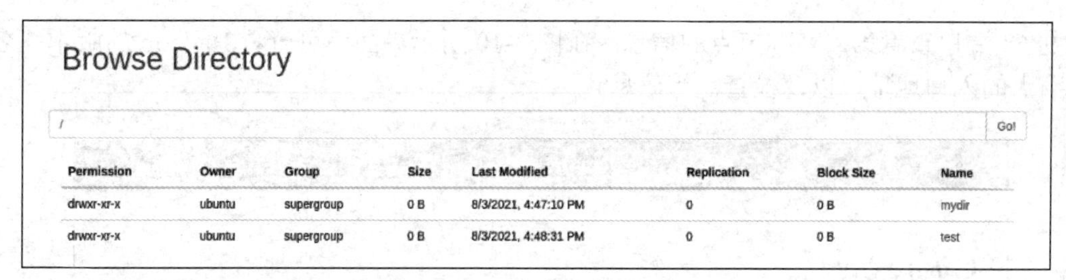

Browse Directory

/ Go!

Permission	Owner	Group	Size	Last Modified	Replication	Block Size	Name
drwxr-xr-x	ubuntu	supergroup	0 B	8/3/2021, 4:47:10 PM	0	0 B	mydir
drwxr-xr-x	ubuntu	supergroup	0 B	8/3/2021, 4:48:31 PM	0	0 B	test

图 3-7　HDFS 的目录结构及文件属性等信息

在图 3-6 所示界面上单击“Datanodes”选项，可以查看文件系统的数据节点信息，如图 3-8 所示。

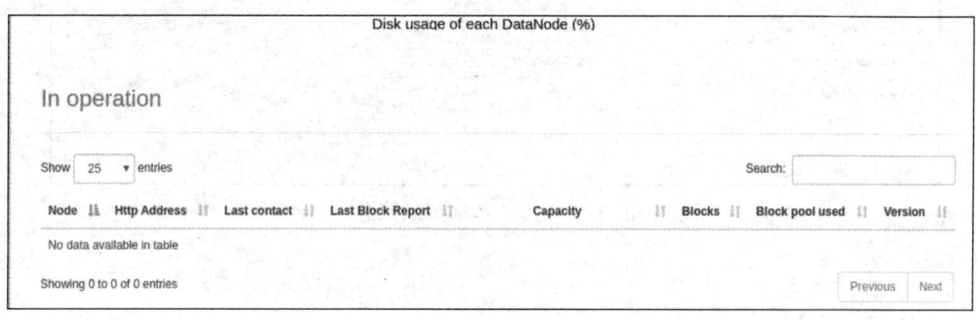

图 3-8　数据节点信息

在图 3-6 所示界面上单击“Startup Progress”选项，可以查看 HDFS 的启动过程，如图 3-9 所示。可以看到，HDFS 的启动经历了 4 个阶段。

Hadoop　Overview　Datanodes　Datanode Volume Failures　Snapshot　Startup Progress　Utilities

Startup Progress

Elapsed Time: 0 sec, Percent Complete: 100%

Phase	Completion	Elapsed Time
Loading fsimage /home/ubuntu/hadoop_data/dfs/name/current/fsimage_0000000000000000000 393 B	**100%**	**0 sec**
erasure coding policies (0/0)	100%	
inodes (1/1)	100%	
delegation tokens (0/0)	100%	
cache pools (0/0)	100%	
Loading edits	**100%**	**0 sec**
Saving checkpoint	**100%**	**0 sec**
Safe mode	**100%**	**0 sec**
awaiting reported blocks (0/0)	100%	

图 3-9　HDFS 的启动过程

在 Linux 操作系统自带的浏览器中输入 http://localhost:9868（内部网址）后回车，通过 9868 端口查看第二名称节点的信息。在图 3-10 所示界面中可查看 Hadoop 的版本、名称节点的入口地址，以及检查点等信息。

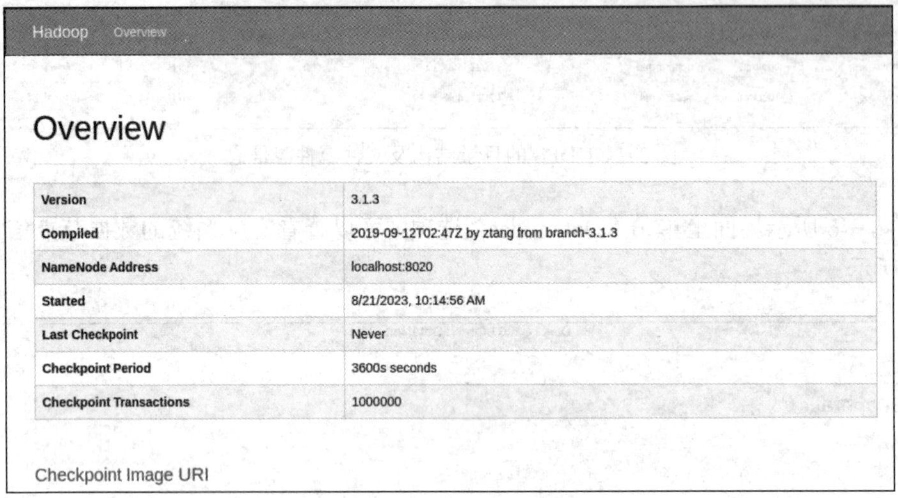

Hadoop Overview

Overview

Version	3.1.3
Compiled	2019-09-12T02:47Z by ztang from branch-3.1.3
NameNode Address	localhost:8020
Started	8/21/2023, 10:14:56 AM
Last Checkpoint	Never
Checkpoint Period	3600s seconds
Checkpoint Transactions	1000000

Checkpoint Image URI

图 3-10　第二名称节点信息

3.1.6　任务实现

首先，把本地计算机硬盘中的数据文件 test.log 传输到集群主服务器（主节点）的本地目录/home/ubuntu/下，可以使用 Xftp 等三方传输工具进行上传。

然后，在主节点的终端执行目录创建和文件上传的命令，具体如下。

```
$hdfs dfs -mkdir -p /user/hadoop/input
$hdfs dfs -put /home/ubuntu/test.log  /user/hadoop/input/test.log
```

最后，检查 HDFS 的/user/hadoop/input/目录，如果该目录下有 test.log 文件，则表示上传成功。

3.2 Java 操作 HDFS

【任务描述】在 Eclipse 中创建与 HDFS 交互的 Java 项目，该项目需实现以下功能：判断 test.txt 文件是否已经存在于 HDFS 的/mydir/input 目录下，如果存在则追加数据“hello, hdfs!”，否则创建文件，并在新建文件中写入数据“hello, hdfs!”。

在实现任务之前，读者需要先掌握 Java 操作 HDFS 的方法，掌握使用 Java API 进行 HDFS 编程的步骤和方法。

Hadoop 采用 Java 语言开发，提供了 Java API 与 HDFS 进行交互的接口。此外，Hadoop 还提供了多种 HDFS 访问接口，例如 C API、HTTP API 和 REST API，其中，C API 为 C 语言程序提供了 HDFS 文件操作和文件系统管理的访问接口，HTTP API 提供了通过 HTTP

远程读取数据的方式。本书主要介绍 Java API，Hadoop 官方网站提供了完整的 Hadoop API 文档，读者可访问该网站进行查看，以进行深入学习。

使用 Java API 进行 HDFS 编程的一般流程如下。

（1）实例化 Configuraion 类。Configuration 类封装了客户端或服务器的配置信息，Configuration 实例会自动加载 HDFS 的配置文件 core-site.xml，并从中获取 Hadoop 集群的配置信息。

（2）实例化 FileSystem 类。FileSystem 类是客户端访问文件系统的入口，是一个抽象的文件系统类。DistributedFileSystem 类是 FileSystem 类的一个具体实现。

（3）设置目标对象的路径。HDFS Java API 提供了 Path 类来封装 HDFS 文件路径，Path 类位于 org.apache.hadoop.fs 包中。

（4）执行文件或目录操作。得到 FileSystem 实例后，我们可以使用该实例提供的方法执行相应的操作，例如打开文件、创建文件、重命名文件、删除文件等。FileSystem 类的常用方法如表 3-4 所示。

表 3-4 FileSystem 类的常用方法

方法名	返回值类型	说明
create(Path f)	FSDataOutputStream	创建一个文件
open(Path f)	FSDataInputStream	打开指定的文件
delete(Path f)	boolean	删除指定文件
exists(Path f)	boolean	检查文件是否存在
getBlockSize(Path f)	long	返回指定文件的数据块的大小
getLength(Path f)	long	返回文件长度
mkdirs(Path f)	boolean	建立子目录
copyFromLocalFile(Path src, Path dst)	void	从本地磁盘上传文件到 HDFS
copyToLocalFile(Path src, Path dst)	void	从 HDFS 下载文件到本地磁盘

下面介绍如何在 Eclipse 中创建与 HDFS 交互的 Java 项目。

3.2.1 在 Eclipse 中创建 HDFS 交互 Java 项目

要在 Eclipse 中创建 HDFS 交互 Java 项目，需要先在实验环境中安装可用的 Eclipse 环境（这里不详细介绍 Eclipse 环境的安装过程，读者可自行查阅相关资料）。安装并启动 Eclipse，之后设置工作空间，工作空间的路径可根据实际情况进行指定，也可直接使用默认路径。以下实验中直接选用默认路径/home/ubuntu/eclipse-workspace，如图 3-11 所示。

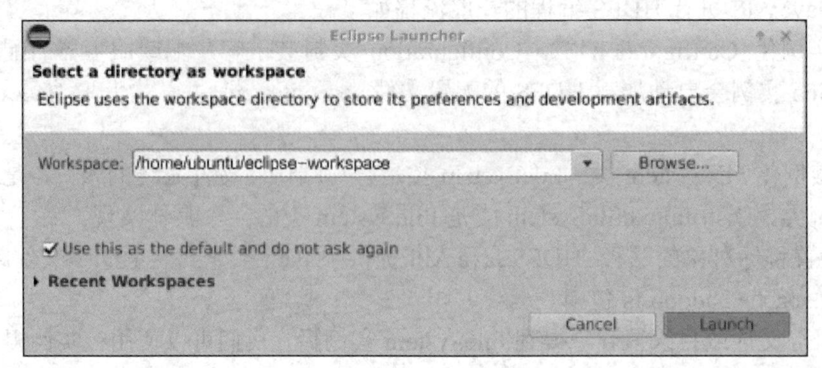

图 3-11　设置工作空间

Eclipse 环境启动完成之后进入工作界面，如图 3-12 所示。

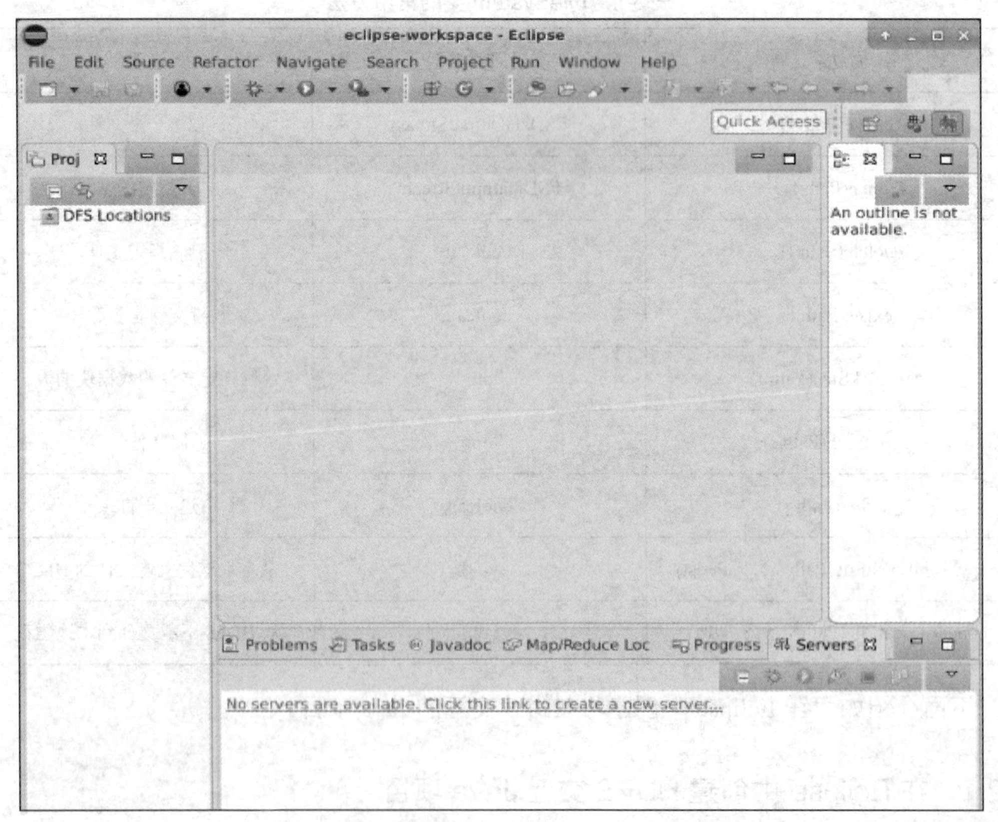

图 3-12　工作界面

我们选择“File”→“New”→“Project”选项，创建一个 Java 项目。创建项目如图 3-13 所示，选择项目类型如图 3-14 所示。

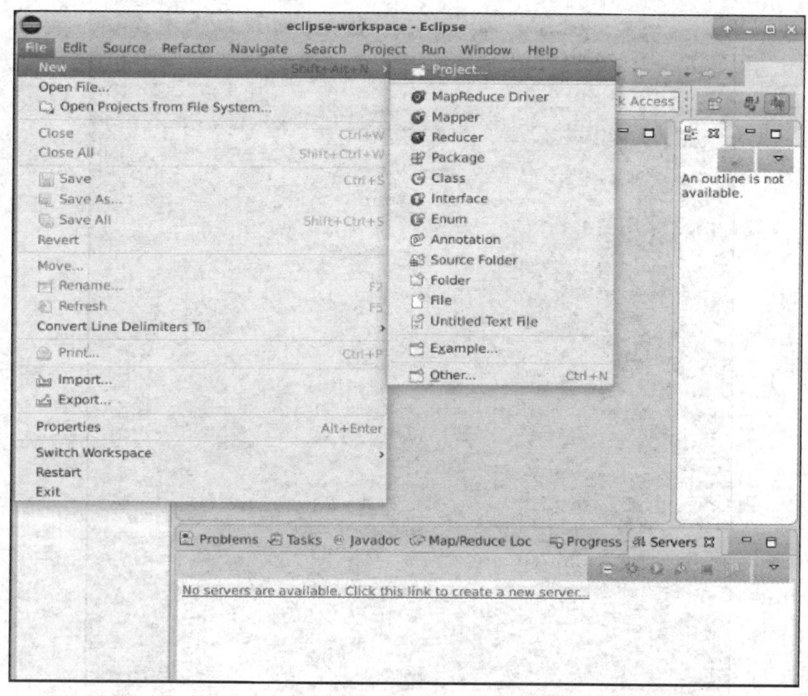

图 3-13 创建项目

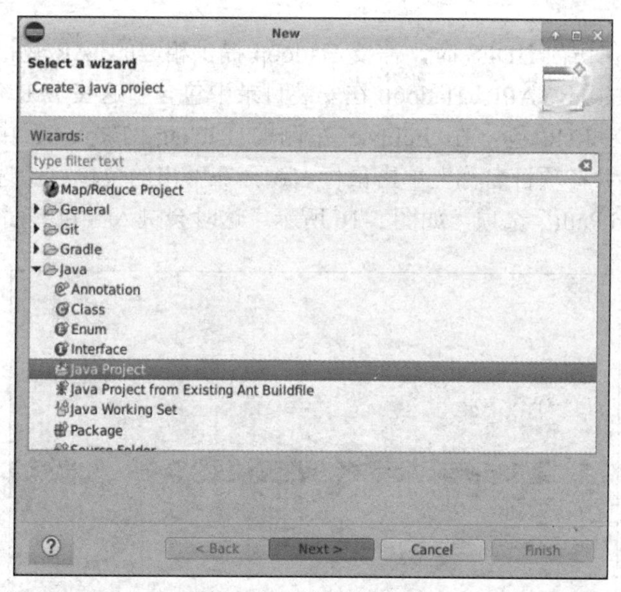

图 3-14 选择项目类型

项目名称及目录设置如图 3-15 所示，在“Project name”文本框中填写项目名称 HDFSDemo，选中“Use default location”复选框，将 Java 工程的所有文件都保存在 /home/unbuntu/eclipse-workspace/HDFSDemo 目录下。之后，在 JRE 选项卡中选择当前系统已安装好的 JDK，并单击“Finish”按钮完成项目的创建。

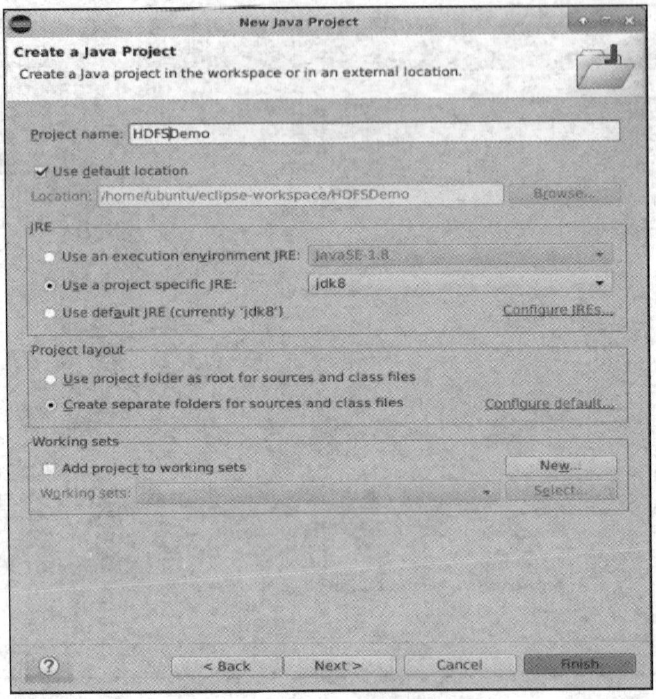

图 3-15　项目名称及目录设置

Java 应用程序在访问 HDFS 时，需要 Hadoop 提供相应的 JAR 包，这些 JAR 包包含了可以访问 HDFS 的 Java API。Hadoop 在安装目录中包含了这些 JAR 包，接下来我们为 Java 项目添加这些 JAR 包。在 Eclipse 左侧的 Package Explorer 中找到刚才创建的 HDFSDemo 项目，并在项目名称上单击鼠标右键，在弹出的菜单栏中选择“Build Path”→“Configure Build Path”选项，如图 3-16 所示，这时会进入导包界面。

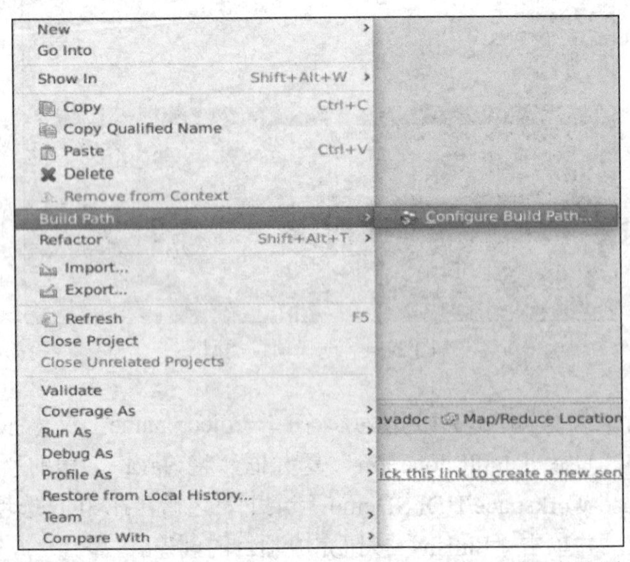

图 3-16　选择“Build Path”→“Configure Build Path”选项

在图 3-17 所示的导包界面中，先单击“Libraries”选项卡，然后单击界面右侧的“Add External JARs”按钮，在弹出的界面中选择所有需要的 JAR 包文件后，单击“Apply and Close”按钮，完成 JAR 包的导入。

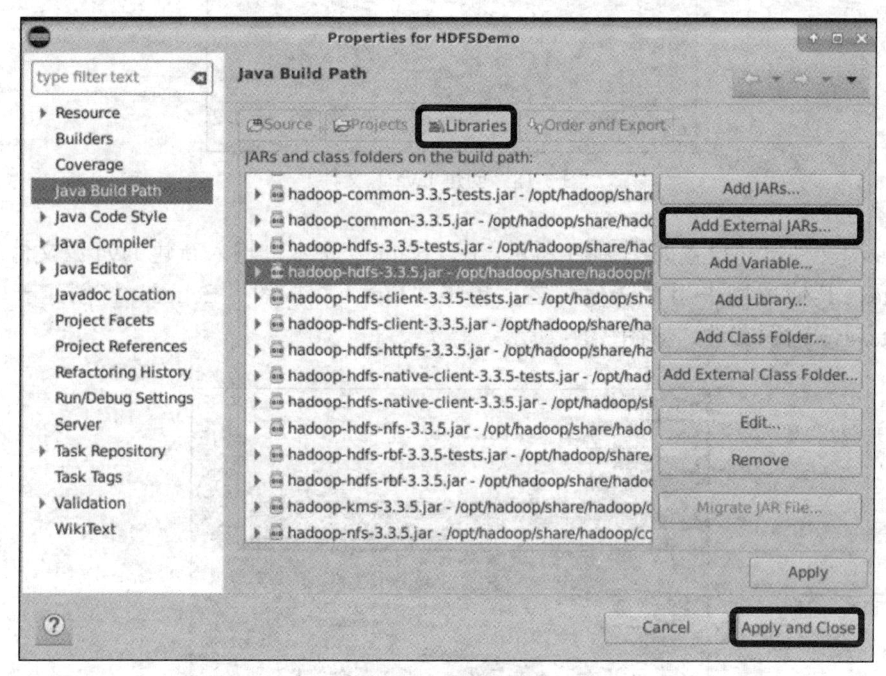

图 3-17　导包界面

要编写一个能够与 HDFS 交互的 Java 项目，需先为项目导入以下 JAR 包。

- /opt/hadoop/share/hadoop/common 目录下的 hadoop-common-3.3.5.jar 和 hadoop-nfs-3.3.5.jar。
- /opt/hadoop/share/hadoop/common/lib 目录下的所有 JAR 包。
- /opt/hadoop/share/hadoop/hdfs 目录下的 hadoop-hdfs-3.3.5.jar 和 hadoop-hdfs-nfd-3.3.5.jar。
- /opt /hadoop/share/hadoop/hdfs/lib 目录下的所有 JAR 包。

请注意，上面的路径中的/opt /hadoop 表示 Hadoop 的安装目录，因此，读者在实际操作时需要根据 Hadoop 实际安装路径选择对应路径下的 JAR 包。JAR 包版本号也需根据实际情况进行选择，无须和上面的版本号保持完全一致。

3.2.2　在 Java 项目中编写 Java 应用程序

我们以在 HDFS 上创建文件为例，讲解与 HDFS 交互的 Java 应用程序的编写方法。

在 Eclipse 左侧的 Package Explorer 中找到刚才创建的项目 HDFSDemo，在项目名称上单击鼠标右键，并在弹出的菜单中选择“New”→“Class”选项，创建 Java 源文件，如图 3-18 所示。

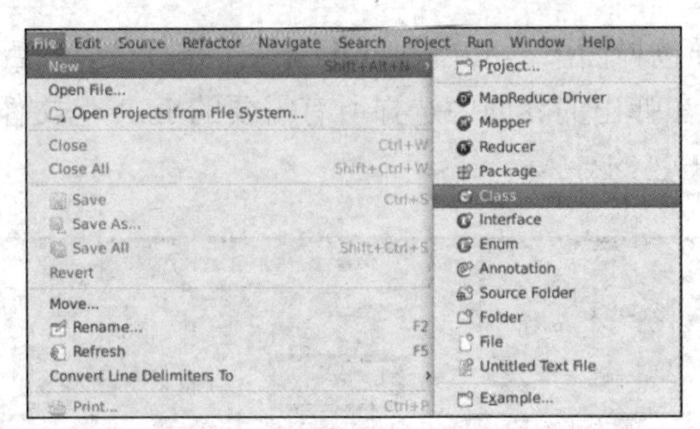

图 3-18　创建 Java 源文件

Java 源文件属性设置如图 3-19 所示。在“Name”文本框输入新建 Java 源文件的名称 fileCreate，其他项均采用默认值，之后单击“Finish”按钮。Eclipse 将创建一个名为 fileCreate.java 的源文件，如图 3-20 所示。

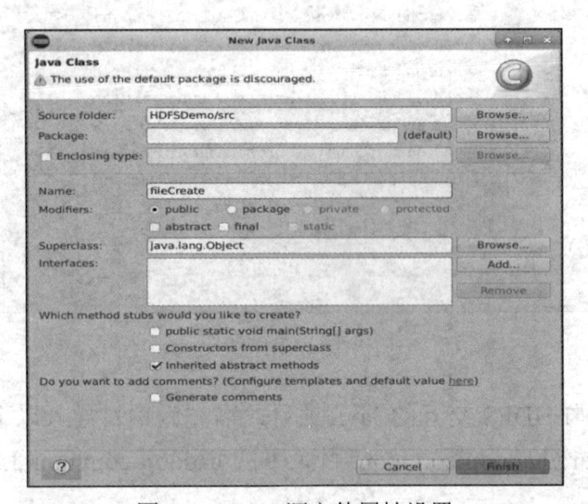

图 3-19　Java 源文件属性设置

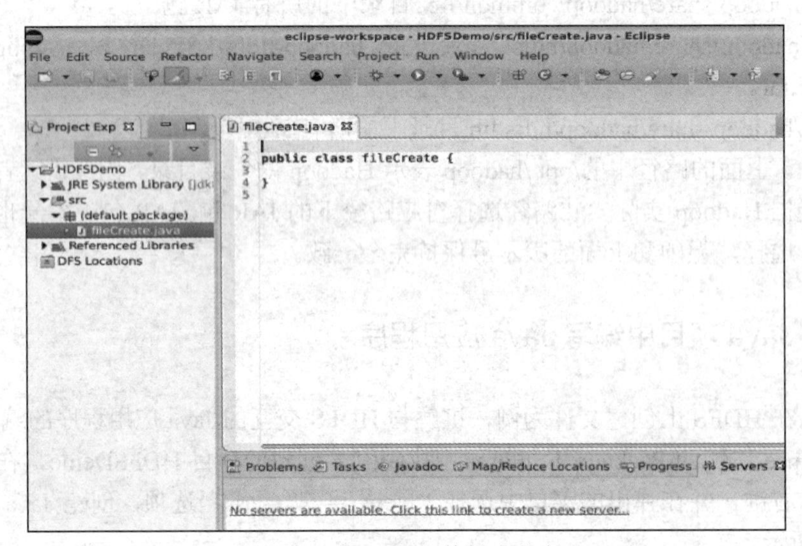

图 3-20　Java 源文件创建成功

我们在 fileCreate.java 源文件中输入以下内容。这段代码能在 HDFS 的/mydir 目录下创建一个名为 test.txt 的文件，该文件包含“hello, hdfs!”。

```
/**
*在 HDFS 上创建/mydir/test.txt 文件
*/
import org.apache.hadoop.conf.*;
import org.apache.hadoop.fs.*;
import java.net.URI;
public class fileCreate{
    public static void main(String[] args) throws Exception{
        // 获取文件系统
        Configuration conf = new Configuration();
        // 配置名称节点地址
        URI uri = new URI("hdfs:// 192.168.30.121:8020");
        // 指定用户名，获取 FileSystem 对象
        FileSystem fs = FileSystem.get(uri,conf,"hadoop");
        // define new file
        Path dfs = new Path("/mydir/test.txt");
        FSDataOutputStream os = fs.create(dfs,true);
        newFile.writeBytes("hello,hdfs!");
        // 关闭数据流
        os.close();
        // 不需要再操作 FileSystem 对象，关闭 FileSystem 对象
        fs.close();
    }
}
```

3.2.3　编译运行应用程序与打包文件

1．编译运行应用程序

运行程序之前需要先启动 HDFS，即在 HDFS 部署环境中打开一个终端，输入 start-dfs.sh 或 start-all.sh 命令，如图 3-21 所示。启动 HDFS 后，我们使用 jps 命令查看 HDFS 的所有进程是否正常运行，确保 HDFS 可正常运行。

```
ubuntu@d28f4a570431:~$ start-dfs.sh
Starting namenodes on [localhost]
localhost: starting namenode, logging to /opt
/hadoop/logs/hadoop-ubuntu-namenode-d28f4a570
431.out
localhost: starting datanode, logging to /opt
/hadoop/logs/hadoop-ubuntu-datanode-d28f4a570
431.out
Starting secondary namenodes [0.0.0.0]
0.0.0.0: starting secondarynamenode, logging
to /opt/hadoop/logs/hadoop-ubuntu-secondaryna
menode-d28f4a570431.out
ubuntu@d28f4a570431:~$ jps
1905 Jps
339 org.eclipse.equinox.launcher_1.4.0.v20161
219-1356.jar
1477 NameNode
1606 DataNode
1773 SecondaryNameNode
ubuntu@d28f4a570431:~$
```

图 3-21　启动 HDFS

我们回到 Eclipse 工作界面，在 fileCreate.java 源文件编辑界面空白位置单击鼠标右键，在弹出的菜单中选择“Run As”→“Java Application”选项，运行 Java 应用程序，如图 3-22 所示。

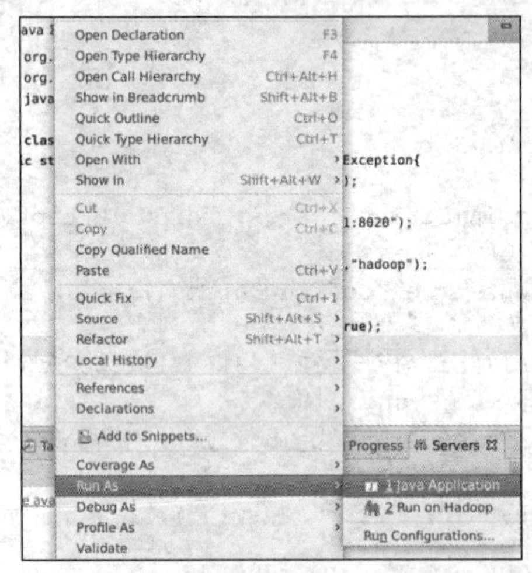

图 3-22　运行 Java 应用程序

在 HDFS 部署环境中的终端输入以下命令，查看新创建的 test.txt 文件及其内容。

```
$hdfs dfs -cat /mydir/test.txt
```

结果如下。

```
Hello,hdfs!
```

2. 打包文件

应用程序源文件编写好后，可以被打包成 JAR 包，方便后续调用。下面介绍应用程序源文件如何打包成 JAR 包。

在 Eclipse 界面左侧的 Package Explorer 中找到需要打包的项目 HDFSDemo，在项目名称上单击鼠标右键，在弹出的菜单中选择“Export”选项，如图 3-23 所示。

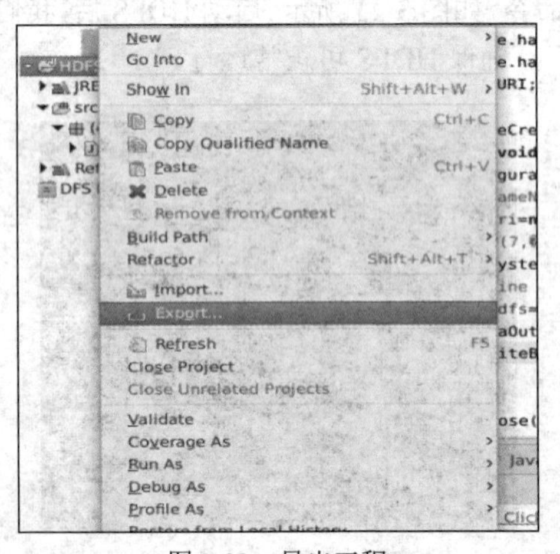

图 3-23　导出工程

在图 3-24 所示的 Export（导出）界面中，选择“JAR file”选项，表示将文件输出为 JAR 包。这时弹出图 3-25 所示对话框，在该界面上选择对哪些文件进行打包，并在“JAR file”输入框中输入生成的 JAR 包的存储路径及包名称，如图 3-25 和图 3-26 所示。

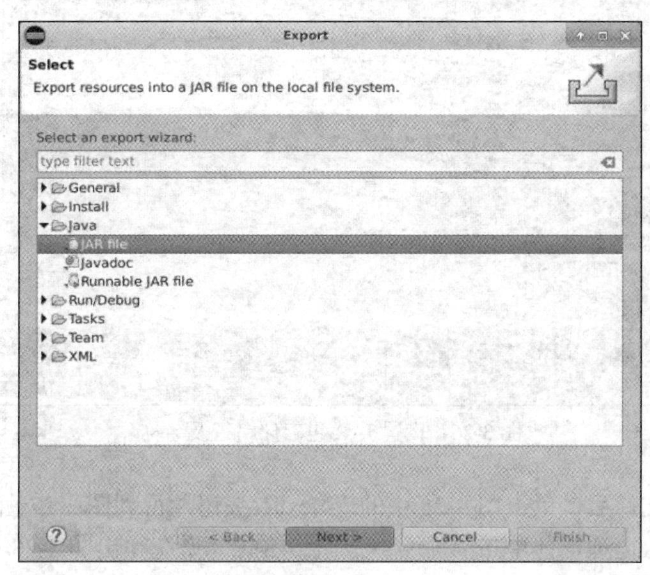

图 3-24　导出界面

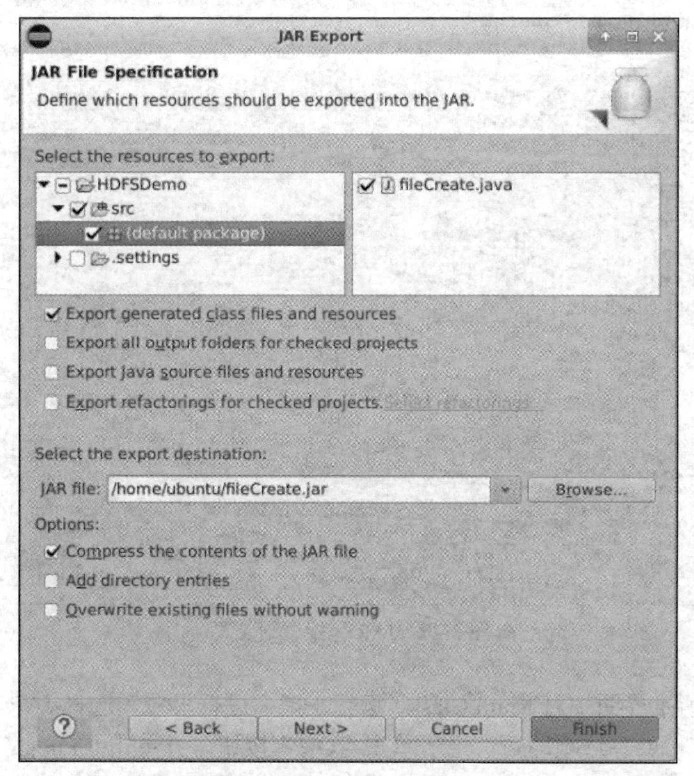

图 3-25　选择需打包的文件

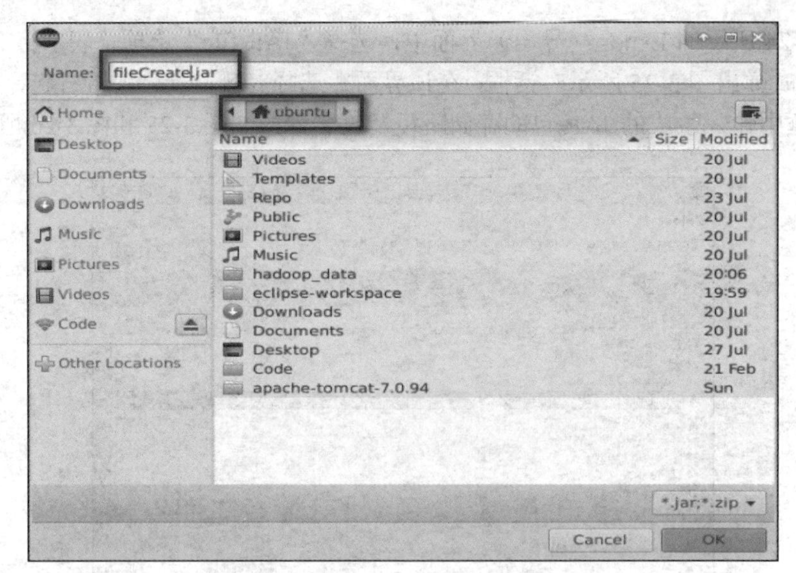

图 3-26　设置 JAR 包存储路径及名称

选择好打包文件后，单击图 3-25 中的“Next”按钮，得到图 3-27 所示对话框。在该对话框中选择默认选项，并单击“Next”按钮。这时，系统弹出图 3-28 所示界面。

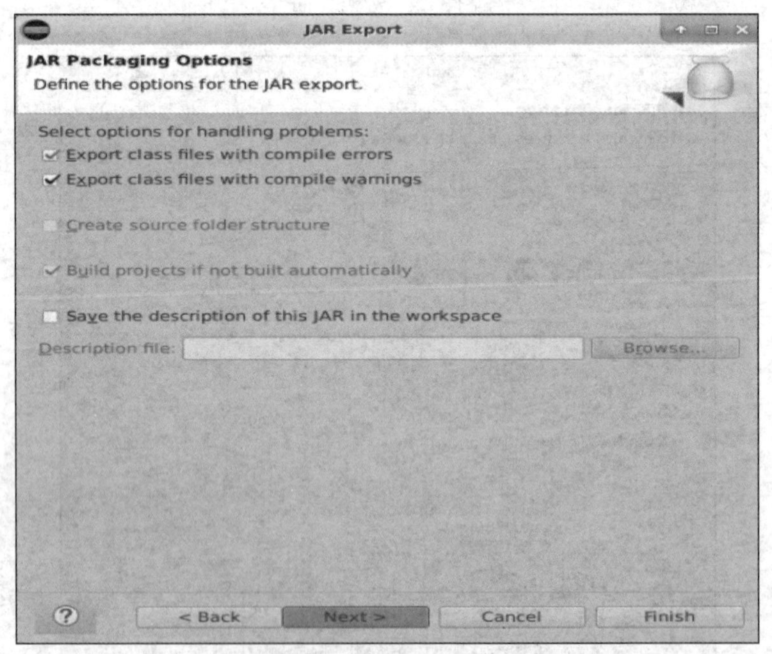

图 3-27　JAR 包配置对话框

在图 3-28 所示界面中对 Main Class（运行主类）进行设置，选择 JAR 包的运行主类。此处选择“fileCreate”选项即可，之后单击“OK”按钮，并单击“Finish”按钮，完成应用程序源文件的打包。

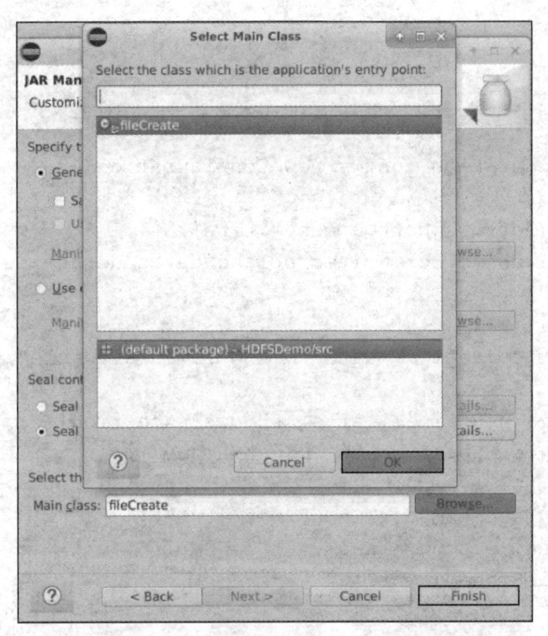

图 3-28 设置 JAR 包的运行主类

我们对 JAR 包进行测试。首先将刚打包好的 JAR 包移动到 HDFS 部署环境的/opt/hadoop/testapp 目录下（其中的/opt /hadoop/目录为 Hadoop 安装目录，testapp 目录需在/opt /hadoop/目录下手动创建），并将之前创建的/mydir/test.txt 文件删除，然后在终端输入以下命令。

```
$cd /opt/hadoop
$./bin/hadoop jar ./testapp/fileCreate.jar
```

上述命令执行完后，我们使用以下命令查看这次新创建的文件及其内容。

```
$hdfs dfs -cat /mydir/test.txt
```

结果如下。

```
Hello,hdfs!
```

3.2.4 任务实现

Java 操作 HDFS 任务实现的具体步骤如下。

步骤 1：打开 Eclipse，创建一个新项目 TestDemo，并导入 Java 项目与 HDFS 交互所需要的 JAR 包。

步骤 2：创建 Java 源文件 Test.java，编写 Java 应用程序代码。代码的具体内容如下，可实现以下功能：判断文件是否已经存在于 HDFS，如果存在则追加数据，如果不存在则创建文件，并在新创建的文件中写入数据。

```
/**
*判断文件是否已经存在于 HDFS，如果存在则追加数据，如果不存在则创建文件，并在新创建的文件中写入
 数据
*/
import org.apache.hadoop.conf.*;
```

```
import org.apache.hadoop.fs.*;
import java.net.URI;

public class Test {
private static Configuration conf = new Configuration();
      static {
    conf.setBoolean("dfs.support.append",true);
    conf.set("dfs.client.block.write.replace-datanode-on-failure.policy",
            "NEVER");
    conf.setBoolean(
"dfs.client.block.write.replace-datanode-on-failure.enable",true);
}
public static void main(String[] args) throws Exception{
    ObjectMapper objectMapper = new ObjectMapper();
    FileSystem fs = null;
    Path path = new Path("/mydir/input/test.txt");
    FSDataOutputStream output = null;
    fs = path.getFileSystem(conf);
    // 如果此文件不存在则创建新文件
    if (!fs.exists(path)) {
        fs.createNewFile(path);
        }
        output = fs.append(path);
        System.out.println("hello,hdfs!".toString());
        output.write(objectMapper.writeValueAsString("hello,hdfs!").getBytes
                ("UTF-8"));
        output.write("\n".getBytes("UTF-8"));// 换行
        fs.close();
        output.close();
    }
}
```

步骤 3：在集群主服务器的终端上执行 jps 命令，查看 HDFS 所有进程是否正常运行。如果没启动 HDFS，则先启动 HDFS，以保证 HDFS 的所有进程正常运行。

步骤 4：在集群主服务器的终端上执行查看命令，查看 HDFS 是否存在/mydir/input/test.txt 文件，如果存在，则删除 test.txt 文件。

步骤 5：在 fileCreate.java 源文件编辑界面的空白位置上单击鼠标右键，在弹出的菜单中选择“Run As”→“Java Application”选项，运行程序。如果在运行代码的过程中，系统报错，则根据报错信息修改代码，直到正常执行完所有代码为止。

步骤 6：在集群主服务器的终端上执行查看命令，查看 HDFS 是否存在/mydir/input/test.txt 文件，若存在则打开 test.txt 文件，查看其中是否包含数据“hello, hdfs!”。如果不存在 test.txt 文件或文件中无“hello, hdfs!”，则修改代码，以创建 test.txt 文件或增加“hello, hdfs!”数据。

步骤 7：再次运行 Java 应用程序代码，重复执行一次步骤 5。

步骤 8：在集群主服务器的终端上执行查看命令，查看/mydir/input/test.txt 文件是否已包含数据“hello, hdfs!”，如果已包含则说明任务完成，反之则任务未完成，需继续修改代码，直至成功。

3.2.5 文件常用操作的参考代码

（1）创建文件

通过 FileSystem.create(Path f, Boolean b)方法在 HDFS 上创建文件，参考代码如下。

```
/**
*在 HDFS 上创建/mydir/test.txt 文件
*/
import org.apache.hadoop.conf.*;
import org.apache.hadoop.fs.*;
import java.net.URI;

/**
 * HDFS:  Create File
 *
 */
public class App {
    public static void main(String[] args) throws Exception{
        Configuration conf = new Configuration();
        // 配置名称节点地址
        URI uri = new URI("hdfs:// 192.168.30.121:8020");
        // 指定用户名，获取 FileSystem 对象
        FileSystem fs = FileSystem.get(uri,conf,"hadoop");
        // define new file
        Path dfs = new Path("/mydir/test.txt");
        FSDataOutputStream os = fs.create(dfs,true);
        os.writeBytes("hello,hdfs!");

        // 关闭流
        os.close();
       // 不需要再操作 FileSystem 对象，关闭 FileSystem 对象
       fs.close();
    }
}
```

通过 HDFS 命令查看结果，参考代码如下。

```
hdfs dfs -cat /mydir/test.txt
```

结果如下。

```
Hello,hdfs!
```

（2）上传文件

文件上传可以采用以下两种方式。

方式 1：采用 FileSystem 类自带的 copyFromLocalFile 接口上传文件，参考代码如下。

```
/**
*将本地 E:\data.txt 文件上传至 HDFS 的/mydir 目录下
*/
import java.io.FileInputStream;
import java.io.InputStream;
```

```
import java.io.OutputStream;
import java.net.URI;

import org.apache.hadoop.conf.Configuration;
import org.apache.hadoop.fs.FileSystem;
import org.apache.hadoop.fs.Path;
import org.apache.hadoop.io.IOUtils;

public class App
{
    public static void main(String[] args) throws Exception{
        Configuration conf = new Configuration();
        // 配置名称节点地址
        URI uri = new URI("hdfs:// 192.168.30.121:8020");
        // 指定用户名，获取 FileSystem 对象
        FileSystem fs = FileSystem.get(uri,conf,"hadoop");

        // 本地文件
        Path src = new Path("e:\\data.txt");
        // HDFS 文件
        Path dst = new Path("/mydir/data.txt");
        fs.copyFromLocalFile(src,dst);

        // 不需要再操作 FileSystem 对象，关闭 FileSystem 对象
        fs.close();
        System.out.println( "upload Successfully!" );
    }
}
```

方式 2：采用流拷贝的方式上传文件，参考代码如下。

```
/**
*将本地 E:\data.txt 文件上传至 HDFS 的/mydir 下
*/
import java.io.FileInputStream;
import java.io.InputStream;
import java.io.OutputStream;
import java.net.URI;
import org.apache.hadoop.conf.Configuration;
import org.apache.hadoop.fs.FileSystem;
import org.apache.hadoop.fs.Path;
import org.apache.hadoop.io.IOUtils;

public class App
{
    public static void main( String[] args ) throws Exception {
        Configuration conf = new Configuration();
        // 配置名称节点地址
        URI uri = new URI("hdfs:// 192.168.30.121:8020");
        // 指定用户名，获取 FileSystem 对象
        FileSystem fs = FileSystem.get(uri,conf,"hadoop");
```

```
        // 构造一个输入流
        InputStream is = new FileInputStream("e:\\data.txt");

        // 得到一个输出流
        OutputStream os = fs.create(new Path("/mydir/data.txt"));

        // 使用工具类实现复制
        IOUtils.copyBytes(is,os,1024);

        // 关闭流
        is.close();
        os.close();

        // 不需要再操作 FileSystem 对象，关闭 FileSystem 对象
        fs.close();

        System.out.println( "Upload Successfully!" );
    }
}
```

通过 HDFS 命令查看结果，参考代码如下。

```
hdfs dfs -ls /mydir
```

结果如下。

```
Found 1 items
-rw-r--r--   3 hadoop supergroup   27 2023-08-02 09:10 /mydir/data.txt
```

（3）下载文件

文件下载可以通过以下两种方式。

方式 1：采用 FileSystem 类自带的 copyToLocalFile 接口下载文件，参考代码如下。

```
/**
*将 HDFS 上的/mydir/test.txt 文件下载到本地 E 盘根目录下
*/
import java.net.URI;
import org.apache.hadoop.conf.Configuration;
import org.apache.hadoop.fs.FileSystem;
import org.apache.hadoop.fs.Path;

public class App
{
    public static void main(String[] args) throws Exception{
        Configuration conf = new Configuration();
        // 配置名称节点地址
        URI uri = new URI("hdfs:// 192.168.30.121:8020");

        // 指定用户名，获取 FileSystem 对象
        FileSystem fs = FileSystem.get(uri,conf,"hadoop");

        // HDFS file
```

```
        Path src = new Path("/mydir/test.txt");
        // local file
        Path dst = new Path("e:\\test.txt");

        // Linux 操作系统下
        // fs.copyToLocalFile(src,dst);

        // Windows 下
        fs.copyToLocalFile(false,src,dst,true);

        // 不需要再操作 FileSystem 对象，关闭 FileSystem 对象
        fs.close();
        System.out.println( "Download Successfully!" );
    }
}
```

方式 2：采用流拷贝的方式下载文件，参考代码如下。

```
import java.io.FileOutputStream;
import java.io.InputStream;
import java.io.OutputStream;
import java.net.URI;
import org.apache.hadoop.conf.Configuration;
import org.apache.hadoop.fs.FileSystem;
import org.apache.hadoop.fs.Path;
import org.apache.hadoop.io.IOUtils;

public class App
{
    public static void main( String[] args ) throws Exception
    {
        Configuration conf = new Configuration();
        // 配置名称节点地址
        URI uri = new URI("hdfs:// 192.168.30.121:8020");
        // 指定用户名，获取 FileSystem 对象
        FileSystem client = FileSystem.get(uri,conf,"hadoop");

        // 打开一个输入流 <------HDFS
        InputStream is = client.open(new Path("/mydir/test.txt"));

        // 构造一个输出流  ----> e:\test.txt
        OutputStream os = new FileOutputStream("e:\\test.txt");

        // 使用工具类实现复制
        IOUtils.copyBytes(is, os, 1024);

        // 关闭流
        is.close();
        os.close();

        // 不需要再操作 FileSystem 对象，关闭 client
```

```
        client.close();
        System.out.println( "Download Successfully!" );
    }
}
```

运行后，查看本地目录 E 盘，如果多了一个 test.txt 文件，说明下载成功。

（4）查看文件详细信息

通过 FileStatus 类查看指定文件在 HDFS 集群上的详细信息，例如最近访问时间、最后修改时间、文件大小、数据块大小等，参考代码如下。

```
/**
* 查看 HDFS 上的/mydir/data.txt 文件的详细信息
*/

import java.net.URI;
import java.text.SimpleDateFormat;
import java.util.Date;
import org.apache.hadoop.conf.Configuration;
import org.apache.hadoop.fs.FileStatus;
import org.apache.hadoop.fs.FileSystem;
import org.apache.hadoop.fs.Path;

public class App
{
    public static void main(String[] args) throws Exception {
        Configuration conf = new Configuration();
        // 配置名称节点地址
        URI uri = new URI("hdfs:// 192.168.30.121:8020");
        // 指定用户名，获取 FileSystem 对象
        FileSystem fs = FileSystem.get(uri,conf,"hadoop");

        // 指定路径
        Path path = new Path("/mydir/data.txt");

        // 获取状态
        FileStatus fileStatus = fs.getFileLinkStatus(path);

        // 获取数据块大小
        long blockSize = fileStatus.getBlockSize();
        System.out.println("blockSize:"+blockSize);

        // 获取文件大小
        long fileSize = fileStatus.getLen();
        System.out.println("fileSize:"+fileSize);

        // 获取文件拥有者
        String fileOwner = fileStatus.getOwner();
        System.out.println("fileOwner:"+fileOwner);

        // 获取最近访问时间
```

```
        SimpleDateFormat sdf = new SimpleDateFormat("yyyy-mm-dd hh:mm:ss");
        long accessTime = fileStatus.getAccessTime();
        System.out.println("accessTime:"+sdf.format(new Date(accessTime)));

        // 获取最后修改时间
        long modifyTime = fileStatus.getModificationTime();
        System.out.println("modifyTime:"+sdf.format(new Date(modifyTime)));

        // 不需要再操作 FileSystem 对象，关闭 FileSystem 对象
        fs.close();
    }
}
```

结果如下。

```
blockSize:134217728
fileSize:27
fileOwner:hadoop
accessTime:2023-08-02 10:10:04
modifyTime:2023-02-02 10:10:04
```

（5）删除文件

调用 delete(Path f)方法，从 HDFS 中删除指定文件，参考代码如下。

```
/**
*删除 HDFS 上的/mydir/test.txt 文件
*/
import java.net.URI;
import org.apache.hadoop.conf.Configuration;
import org.apache.hadoop.fs.FileSystem;
import org.apache.hadoop.fs.Path;

public class App{
    public static void main(String[] args) throws Exception {
    Configuration conf = new Configuration();
    // 配置名称节点地址
    URI uri = new URI("hdfs:// 192.168.30.121:8020");
    // 指定用户名，获取 FileSystem 对象
    FileSystem fs = FileSystem.get(uri,conf,"hadoop");

    // HDFS file
    Path path = new Path("/mydir/test.txt");
    fs.delete(path);

    // 不需要再操作 FileSystem 了，关闭
    fs.close();

    System.out.println( "Delete File Successfully!" );
    }
}
```

通过 HDFS 命令查看结果，参考代码如下。

```
hdfs dfs -ls /mydir
```

如果看到/mydir/test.txt 文件已经不存在，说明删除成功。

3.3 本章小结

本章首先介绍 HDFS 的架构、读/写流程；然后详细介绍 HDFS Shell 命令、Web 界面管理 HDFS，以及 Java 操作 HDFS 的应用程序编写与运行方法。通过本章的学习，我们希望读者对 HDFS 有一个整体的认识，了解 HDFS 的架构和组件功能，了解 HDFS 的读/写机制，熟练掌握 HDFS 的访问方式，能熟练使用相关命令处理实际问题。

第 4 章 MapReduce 基本原理与编程实现

MapReduce 是一种编程模型，用于大规模数据集的并行运算。它将复杂的、运行于大规模集群上的并行计算过程高度抽象为两个函数——Map（映射）函数和 Reduce（归约）函数，这极大地方便了分布式编程工作。开发人员在不熟悉分布式并行编程的情况下，也可以将自己的程序运行在分布式系统上，完成海量数据的计算。

【学习目标】

1．掌握 MapReduce 的基本原理。

2．编程实现按日期统计访问次数及按访问次数排序功能。

4.1 MapReduce 基本原理

【任务描述】熟悉 MapReduce 的编程思想，掌握 MapReduce 的相关操作。

4.1.1 MapReduce 简介

MapReduce 是一个分布式运算程序的编程框架，是用户开发基于 Hadoop 的数据分析应用的核心框架。

MapReduce 的核心功能是将用户编写的业务逻辑代码和自带的默认组件整合成一个完整的分布式运算程序，并发运行在一个 Hadoop 集群上。

MapReduce 采用“分而治之”的思想，把对大规模数据集的操作分发给一个主节点管理下的各个子节点，然后整合各个子节点的中间结果，得到最终的计算结果。简而言之，MapReduce 就是“分散任务，汇总结果”。

MapReduce 的优点如下。

① 易于编程。利用 MapReduce 的一些简单接口就可以完成一个分布式程序，这个分布式程序可以运行在大量低配的计算机上。

② 良好的扩展性。当计算资源不能得到满足时，可以通过简单地增加计算机来提高 MapReduce 的计算能力。

③ 高容错性。MapReduce 的设计初衷是使程序能够部署在低配的计算机上，这就要

求它具有很高的容错性。例如其中一台主机出问题了，MapReduce 可以将其计算任务转移到另一台主机上运行，不至于让这个任务运行失败。这个过程不需要人工干预，完全由 MapReduce 在内部完成。

④ 适合 PB 级以上海量数据的离线处理。MapReduce 可以做到让上千台服务器并发工作，提供数据处理能力。

MapReduce 也有不擅长的地方，主要表现在以下 3 个方面。

① 不擅长实时计算。MapReduce 无法像 MySQL 一样，在毫秒级或秒级时间内返回结果。

② 不擅长流式计算。流式计算的输入数据是动态的，而 MapReduce 的输入数据是静态的。例如，实时计算 Web Server 产生的日志是 MapReduce 不擅长的数据，这是因为 MapReduce 自身的设计特点决定了数据源必须是静态的。

③ 不擅长有向无环图计算。对于多个应用程序存在依赖关系，后一个应用程序的输入为前一个的输出这种情况，MapReduce 并不是不能处理，而是处理完后，每个 MapReduce 作业的输出结果都会被写入磁盘，占用大量资源，从而导致系统性能非常低下。

4.1.2 MapReduce 编程核心思想

从 MapReduce 自身的命名特点可以看出，MapReduce 程序的运行过程一般分成 2 个阶段：Map 阶段和 Reduce 阶段。Map 阶段会有若干个 MapTask 实例，这些实例完全并行运行，互不相关。Reduce 阶段会有若干个 ReduceTask 实例，这些实例也完全并行运行，互不相关，但其数据依赖 Map 阶段所有 MapTask 实例的输出。

MapTask 实例把一个数据集转化成另一个数据集，单独的元素会被拆分成键值对。

ReduceTask 实例把 MapTask 实例的输出作为输入，把这些键值对的数据合并成一个更小的键值对数据集。ReduceTask 实例始终在 MapTask 实例之后执行。

MapReduce 编程的核心思想如图 4-1 所示。用户只需要编写 map()和 reduce()两个函数，便可以完成简单的分布式程序的设计。

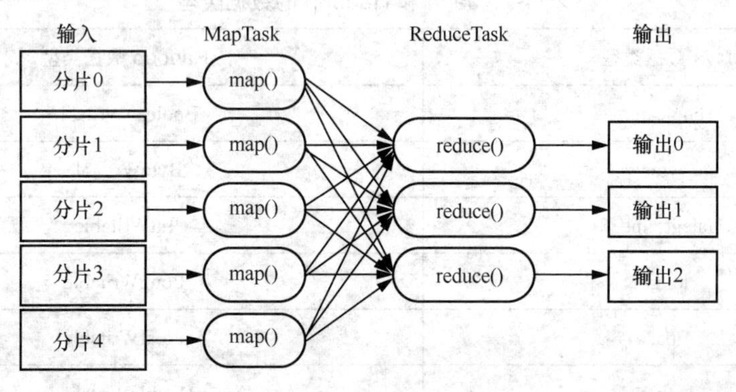

图 4-1 MapReduce 编程的核心思想

MapReduce 编程模型只能包含一个 Map 阶段和一个 Reduce 阶段，如果用户的业务逻辑非常复杂，那么只能串行运行多个 MapReduce 程序。

4.1.3　MapReduce 编程规范

一个 MapReduce 程序可以分成 3 个部分：Mapper、Reducer 和 Driver。MapReduce 编程模型如图 4-2 所示。我们在下文中详细介绍。

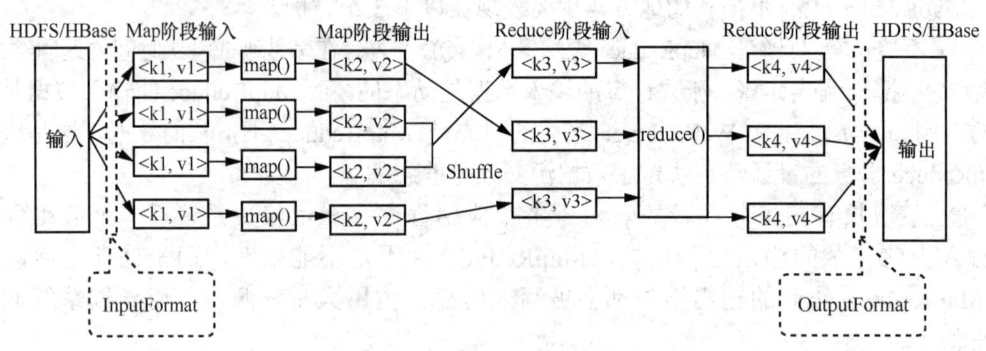

图 4-2　MapReduce 编程模型

图 4-2 体现了 MapReduce 编程模型的几个要点，具体如下。

① 任务 Job = Map + Reduce。

② Map 阶段的输出是 Reduce 阶段的输入。

③ 所有的输入和输出的形式都是键值对。

<k1, v1>是 map()函数的输入，<k2, v2>是 map()函数的输出。

<k3, v3>是 reduce()函数的输入，<k4, v4>是 reduce()函数的输出。

④ k2 = k3，v3 是一个集合，v3 的元素是 v2，表示为 v3 = list(v2)。

MapReduce 编程模型中所有输入和输出的数据类型必须是 Hadoop 的数据类型。Java 和 Hadoop 的数据类型如表 4-1 所示。

表 4-1　Java 和 Hadoop 的数据类型

Java 数据类型	Hadoop 数据类型
boolean	BooleanWritable
byte	ByteWritable
integer/int	IntWritable
long	LongWritable
float	FloatWritable
double	DoubleWritable
string	Text
map	MapWritable

续表

Java 数据类型	Hadoop 数据类型
array	ArrayWritable
null	NullWritable (当<key, value>中的 key 或 value 为空时使用)

4.1.4　MapReduce 的输入格式

1. InputFormat

InputFormat 是 MapReduce 用于处理数据输入的一个组件，是一个抽象父类，主要解决不同数据源的数据输入问题。可以通过下面的语句指定输入格式，如果不指定，默认输入格式为 TextInputFormat。

```
job.setInputFormatClass(TextInputFormat.class);
```

InputFormat 具有以下两个功能。

数据切分：按照某个策略将输入数据切分成若干个 SplitInput（数据分片），以便确定 Mapper 个数及对应的 SplitInput 个数。这样一来，SplitInput 的个数与 Mapper 个数总是相同的。

为 Mapper 提供输入数据：读取给定的 SplitInput 数据，将其解析成键值对，供 Mapper 使用。

InputFormat 处理流程如图 4-3 所示。

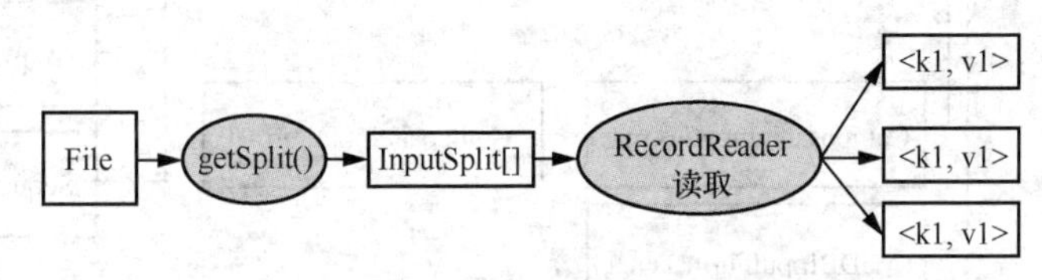

图 4-3　InputFormat 处理流程

InputFormat 有两个重要的方法——getSplit()和 createRecordReader()，这两个方法分别对应上面的两个功能。

```
List<InputSplit> getSplits(JobContext context)
   throws IOException, InterruptedException;
RecordReader<K,V> createRecordReader(InputSplit split,
   TaskAttemptContext context) throws IOException,
   InterruptedException;
```

getSplits()方法负责对数据进行逻辑分片。例如，数据库有 100 条数据（按照 ID 升序存储），假设将 20 条数据分为一片，则分片数是 5。但每个分片只是一种逻辑上的定义，仅提供了如何将数据分片的方法，并没有实现物理上的独立存储。

createRecordReader()方法返回一个 RecordReader 对象，实现了类似迭代器的功能，将某个 InputSplit 解析成键值对。该方法有以下注意事项。

定位记录边界：为了能识别一条完整的记录，需添加一些同步标识，例如 TextInputFormat 的标识是换行符，SequenceFileInputFormat 的标识是每隔若干条记录会添加固定长度的同步字符串。为了避免 InputSplit 中第一条或最后一条出现跨 InputSplit 的情况，createRecordReader()方法规定每个 InputSplit 的第一条不完整记录划给前一个 InputSplit。

解析键值对：将每个记录分解成键和值两部分，TextInputFormat 每行的内容是值，该行在整个文件中的偏移量为键；SequenceFileInputFormat 的记录共有 4 个字段：前两个字段分别是整个记录的长度和键的长度（均为 4B），后两个字段分别是键和值的内容。

2. InputFormat 接口实现类

InputFormat 接口实现类有很多种，其层次结构如图 4-4 所示。

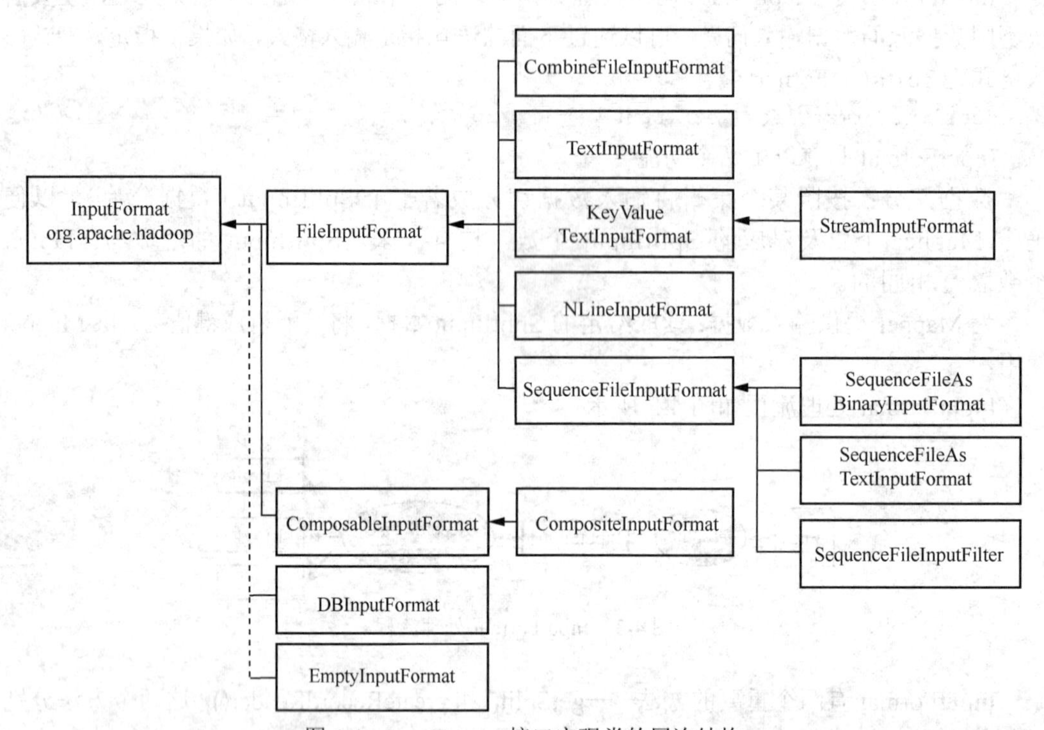

图 4-4　InputFormat 接口实现类的层次结构

这里介绍一下 FileInputFormat。FileInputFormat 是将文件作为数据源的 Input Format 基类，它的主要作用是指出作业的输入文件位置。因为作业的输入被设置为一组路径，这给指定作业的输入提供了很强的灵活性。FileInputFormat 提供了以下 4 种静态方法来设置作业的输入路径。

```
public static void addInputPath(Job job, Path path);
public static void addInputPaths(Job job, String commaSeparatedPaths);
public static void setInputPaths(Job job, String commaSeparatedPaths);
public static void setInputPaths(Job job, Path... inputPaths);
```

4.1.5　MapReduce 的输出格式

针对前面介绍的输入格式，Hadoop 都有相应的输出格式。默认情况下只有一个 Reduce()函数，输出结果只有一个文件，默认文件名为 part-r-00000。输出文件的个数与 Reduce()函数的个数一致。如果有两个 Reduce()函数，则输出结果就有两个文件，第一个为 part-r-00000，第二个为 part-r-00001，依次类推。

1．OutputFormat 接口

OutputFormat 主要用于描述输出数据的格式，能够将用户提供的键值对写入特定格式的文件。我们可以通过 OutputFormat 接口实现具体的输出格式。Hadoop 自带了很多 OutputFormat 接口实现类，它们与 InputFormat 相对应，足够满足日常业务需要。

2．OutputFormat 接口实现类

OutputFormat 接口实现类有很多种，OutputFormat 接口实现类的层次结构如图 4-5 所示。

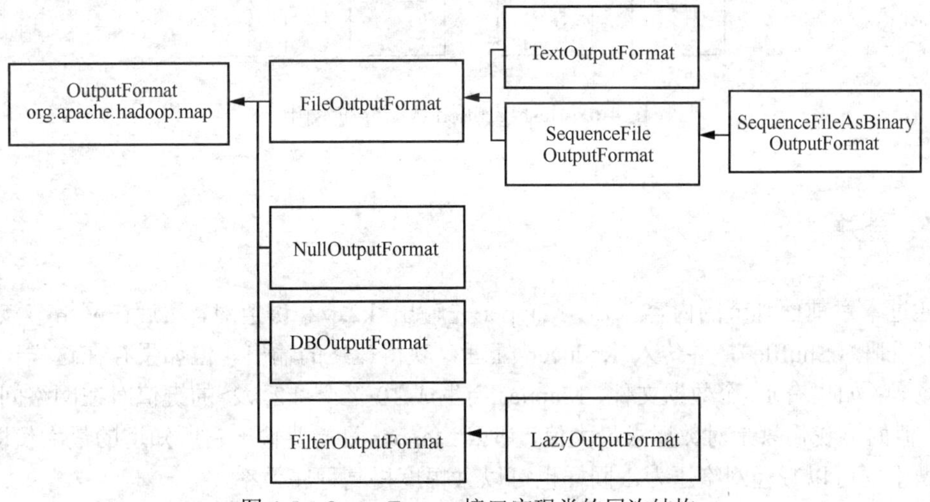

图 4-5　OutputFormat 接口实现类的层次结构

这里介绍一下 TextOutputFormat，它的作用是把每条记录写入文本行。它的键和值可以是实现 Writable 接口的任意类型，因为 TextOutputFormt 调用 toString()方法来把记录转换为字符串。每个键值对由制表符进行分隔，当然也可以设定 mapreduce.output.textoutputformat.separator 属性。

可以使用 NullWritable 来省略输出的键或值（或两者都省略），这个作用相当于 NullOutputFormat 的作用，后者什么都不输出。

4.1.6　分区

在进行 MapReduce 计算时，有时需要把统计结果按照条件输出到不同的文件（分区）

中，例如，将统计结果按照手机归属地不同输出到不同文件（分区）中。我们知道，最终的输出数据来自Reducer，如果要得到多个文件，则意味着有与文件相同数量的ReduceTask在运行。ReduceTask的数据来自MapTask，也就是说Mapper要划分数据，将不同的数据分配给不同的Reducer运行。Mapper划分数据的过程称作分区（Partition），负责实现划分的数据的类称为Partitioner。

MapReduce默认的Partitioner是HashPartitioner。在默认情况下，Partitioner先计算key的散列值（通常为MD5值），然后通过Reducer个数执行求余运算——键的hashCode除以Reducer的个数，取其余数。这种方式不仅能够随机将整个键空间平均分发给每个Reducer，同时也能确保不同Mapper产生的相同键能被分发到同一个Reducer。HashPartition数据分布示例如图4-6所示。

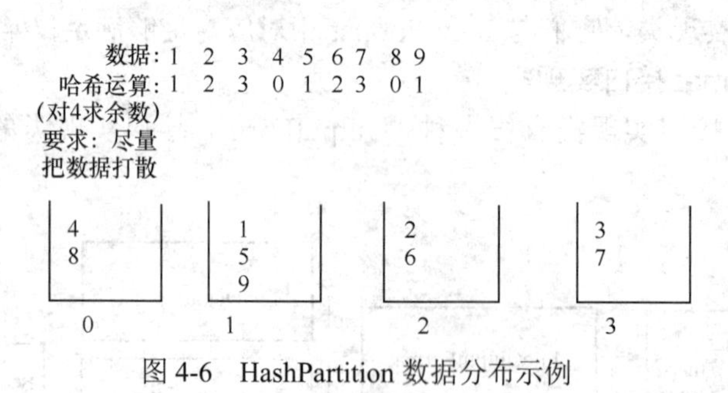

图4-6 HashPartition数据分布示例

4.1.7 合并

通过本章前面几节的内容可知，Mapper先输出<k2, v2>键值对，然后在网络节点间对其进行洗牌（Shuffle），并传入Reducer处理，获得最终的输出。但如果存在这样一个实际的场景：如果有10个数据文件，Mapper会生成10亿个<k2, v2>的键值对在网络间进行传输，但如果我们只是对数据求最大值，那么Mapper只需要输出它所知道的最大值即可。这样做不仅可以减轻网络压力，同样也可以大幅度提高程序效率。

我们可以把合并（Combiner）操作看作在每个单独节点上先做一次Reduce操作，其输入及输出的参数和Reduce是一样的。Combiner组件有以下7个特点。

① Combiner是MR程序中Mapper和Reducer之外的一种组件。

② Combiner组件的父类是Reducer。

③ Combiner和Reducer的区别在于运行的位置，Combiner是在每个MapTask所在的节点上运行，Reducer接收全局所有Mapper的输出结果。

④ Combiner的意义是对每个MapTask的输出进行局部汇总，以减少网络传输量。

⑤ Combiner的应用前提是不能影响最终的业务逻辑，而且Combiner输出的键和值应该与Reducer输入的键和值相对应。

⑥ Combiner可以自定义，自定义时继承一个Reducer类，并重写其中的方法即可。

⑦ Combiner是可以选择的，可以选择不触发，设置后才会触发。

4.2 编程实现——按访问次数排序

【任务描述】首先统计某网站2016年各用户在每个自然日的总访问次数，然后根据访问次数对用户进行排序。原始数据文件中提供了用户名称与访问日期。

这个任务的实质是编写两个MapReduce程序：第一个MapReduce程序用于获取以自然日为单位的各用户的访问次数，并将结果输出在HDFS目录/user/root/AccessCount中；第二个MapReduce程序读取第一个MapReduce程序的结果并进行排序，将排序后的结果存储在HDFS的/user/root/TimesSort目录下。

4.2.1 编程思路与处理逻辑

首先来分析第一个MapReduce程序的编程思路和处理逻辑。进行MapReduce编程时，必须重点考虑以下几个要素。

① 输入/输出格式。

② Mapper要实现的计算逻辑。

③ Reducer要实现的计算逻辑。

（1）定义输入/输出格式

某网站用户的访问日期的格式属于文本格式，访问次数的格式为整型数值格式，访问日期和访问次数组成了键值对为<访问日期,访问次数>。由此可知，Mapper的输出与Reducer的输出分别选用Text类与IntWritable类来实现。

（2）Mapper的逻辑实现

Mapper中最主要的部分是map()函数。map()函数的主要任务是读取原始数据文件中的数据，输出所有访问日期与访问次数的键值对。因为访问日期是数据文件中的第2列，所以先定义一个数组，再提取第2个元素，与初始次数1一起构成要输出的键值对，即<访问日期, 1>。

以伪代码来编写Mapper的处理逻辑，具体如下。

```
Begin
    自定义类 MyMapper 继承自 Mapper;
        覆写 map 函数;
            定义初始次数为 1;
            读取用户访问日志文件;
            以每一行为单位,以逗号为分隔符进行分拆;
            将结果存入 Array 数组;
            将数组中的第 2 个元素与初始次数组合后输出,格式为<访问日期,1>;
End
```

（3）Reducer类的逻辑实现

Reducer类中最主要的部分是reduce函数。reduce函数的主要任务是读取Mapper输出的键值对<访问日期, 1>。该部分的处理逻辑与官方示例WordCount中的Reducer完全相同，在此不做详述。下面以伪代码来表示Reducer的处理逻辑，具体如下。

```
Begin
    自定义MyReducer继承自Reducer;
        覆写reduce函数;
            读取Mapper输出的键估对;
            把相同键的值进行累加;
            输出<访问日期,总访问次数>;
End
```

针对第二个需求分析思路和处理逻辑。首先，对要处理的数据文件进行分析，具体如下。它有两列数据，需要将第 2 列数据按从小到大的顺序对文件中的所有记录进行排序。

```
2016-01-01  5038
2016-01-02  5378
2016-01-03  5341
2016-01-04  5304
2016-01-05  5258
```

前文已经介绍过 MapReduce 原理及执行过程，在此稍做回顾：数据经 Mapper 处理后输出键值对数据，再经过 Shuffle 后传递到 Sorter，Sorter 按 Key 进行排序后，再传递给 Reducer 进行处理，最后的输出结果是按 Key 排好序的。那么正好利用 MapReduce 这个以 Key 排序的特性，对需要排序的数据列进行处理。要特别注意的是，MapReduce 仅对键（Key）进行排序，而不对值（Value）进行排序。所以，必须做相应的数据转换，将要排序的数据列设置为键。通过键值对中的 Key 特性进行排序的处理流程如图 4-7 所示。

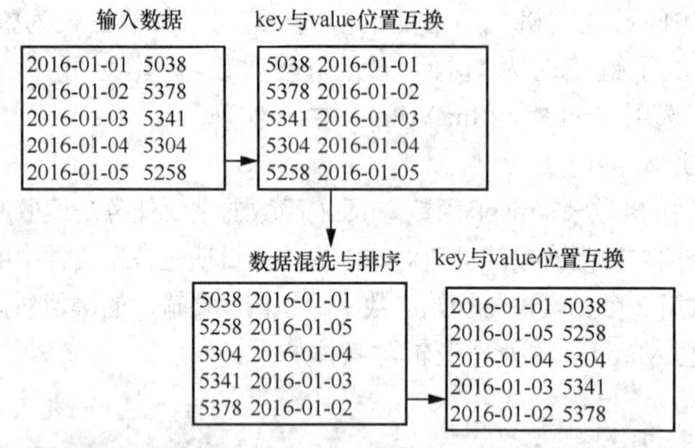

图 4-7　通过键值对中的 Key 特性进行排序的处理流程

Mapper 的处理逻辑相对简单，对于输入的键值对，把 Key 与 Value 位置进行交换，即由<访问日期, 访问次数>转换为<访问次数, 访问日期>。在 Mapper 输出后，键值对经过 Shuttle 与 Sorter 的处理，已按 Key 自动进行了排序，输出格式依然为<访问次数,访问日期>。

Reducer 的处理逻辑与 Mapper 相反，对于输入的键值对，把 Key 与 Value 的位置进行交换，即由<访问次数,访问日期>重新转换为<访问日期,访问次数>。

打开输入与输出格式的选择，日期对应 Text 类，访问次数对应 IntWritable 类，需要根据其在键值对中的具体位置进行配置。

4.2.2　核心模块代码

1. 需求一核心代码

当理解了各个模块的处理逻辑之后，用户就可以编写代码来实现它们。下面以实际代码来编写上一节分析过的几个模块。

① 编写 Mapper 模块代码，具体如下。

```
// Mapper 模块
public static class MyMapper extends Mapper<Object,Text, Text,Intwritable>{
    private final static Intwritable one = new Intwritable(1);
    public void map(Object key,Text value, Context context) throws IOException,
InterruptedException{
        String line = value.toString();
        // 指定逗号为分隔符，组成数组
        String array[] = line.split(", ");
        // 提取数组中的访问日期作为 Key
        String keyoutput = array[1];
        // 组成键值对
        context.write(new Text (keyoutput), one);
    }
}
```

② 编写 Reducer 模块代码，具体如下。

```
// Reducer 模块
public static class MyReducer extends Reducer<Text, Intwritable, Text,
Intwritable>{
    private Intwritable result -new Intwritable ();
    public void reduce (Text key, Iterable<Intwritable> values,Context
context) throws IOException, InterruptedException{
        int sum = 0;// 定义累加器，初始值为 0
        for (IntWritable val : values) {
            // 将相同键的所有值进行累加
            sum += val.get ();
        }
        result.set (sum);
        context.write (key, result);
    }
}
```

③ 编写 Driver 模块，具体如下。

```
public static void main(String[] args) throws Exception{
    Configuration conf = new Configuration();
    Job job=Job.getInstance(conf, "Daily Access Count");
    job.setJarByclass (dailyAccessCount.class);
    ob.setMapperClass (MyMapper.class);
    job.setReducerClass(MyReducer.class);
    job.setMapoutputKeyclass(Text.class);
    job.setMapoutputValueClass(IntWritable.class);
    job.setoutputKeyclass (Text.class);
```

```
    job.setOutputValueClass (Intwritable.class);
    for (int I = 0;i<args.length-1;++i){
        FileInputFormat.addInputPath(job,new Path(args[i]));
    }
    FileoutputFormat.setOutputPath(job,new Path (args[args. length - 1]));
    System.exit(job.waitForCompletion(true) ? 0 : 1);
}
```

2. 需求二核心代码

需求二的任务是对所有数据进行排序，所以需要设置 Reduce 任务的个数为 1，让所有排序结果都输出到同一个文件里。要设置 Reduce 任务数，可在驱动类 Driver 中添加以下代码。

```
job.setNumReduceTasks(num);
```

其中，num 是指 Reduce 任务个数，例如要设置 Reduce 的任务个数为 2，则 nunn 为 2，一般情况下，MapReduce 默认的 Reduce 任务数是 1。因此，如果 Reduce 的任务数为 1，则不必在驱动类中设置 Reduce 任务数。

下面编写核心模块的代码。

① 编写 Mapper 模块代码，实现键值位置交换，具体如下。

```
// Mapper 模块
public static class MyMapper extends Mapper<Object,Text,Intwritable,Text>{
    public void map(Object key,Text value, Context context) throws IOException,
InterruptedException{
        String line = value.toString();
        // 指定 tab 为分隔符，组成数组
        String array[] = line.split("\t ");
        // 提取访问次数作为 Key
        int keyoutput = Integer.parseInt(array[1]);
        // 提取访问日期作为 value
        String valueOutput = array[0];
        // 组成键值对
        context.write(new Intwritable(keyoutput), new Text(valueOutput));
    }
}
```

② 编写 Reducer 模块代码，实现键值位置交换，具体如下。

```
// Reducer 模块
public static class MyReducerextends extends Reducer<Intwritable, Text, Text,
Intwritable>{
    public void reduce (Intwritable key, Iterable<Text> values,Context context)
throws IOException, InterruptedException{
    for (Text value : values){
        context.write(value, key);
    }
}
```

③ 编写 Driver 模块代码，重点是输入键值对与输出键值对的配置，具体如下。

```
public static void main(String[] args) throws Exception{
    Configuration conf = new Configuration ();
    Job job = Job.getInstance (conf, "Access Time Sort");
```

```
    job.setJarByClass(accessTimessort.class);
    job.setMapperClass(MyMapper.class);
    job.setReducerClass(MyReducer.class);
    job.setMapoutputKeyClass(IntWritable.class);
    job.setMapoutputValueClass(Text.class);
    job.setOutputKeyClass(Text.class);
    job.setoutputValueclass (Intwritable.class);
    for (int I = 0;i< args.length - 1;++ i) {
        FileInputFormat.addInputPath(job,new Path(args[i]));
    }
    FileoutputFormat.setoutputPath(job,new Path(args[args. length- 1]));
    System.exit(job.waitForCompletion(true) ? 0 :1);
}
```

4.2.3 任务实现

1. 需求一编写代码来实现用户访问次数统计的处理逻辑

第一部分，分析需求处理逻辑，编写需求处理代码。

① 分析主要模块的处理逻辑，主要是 Mapper 与 Reducer 两个模块。此外，还有输入与输出格式的定义。

② 编写代码实现各个核心模块。

③ 补充与完善代码文件 dailyAccessCount.java，完整内容如下。

```
package test;
import java.io.IOException;
import org.apache.hadoop.conf.Configuration;
import org.apache.hadoop.fs.Path;
import org.apache.hadoop.io.IntWritable;
import org.apache.hadoop.io.Text;
import org.apache.hadoop.mapreduce.Jop;
import org.apache.hadoop.mapreduce.Mapper;
import org.apache.hadoop.mapreduce.Reducer;
import org.apache.hadoop.mapreduce.lib.input.FileInputFormat;
import org.apache.hadoop.mapreduce.lib.output.FileOutputFormat;
import org.apache.hadoop.util.GenericoptionsParser;

public class dailyAccessCount {
    public static class MyMapper extends Mapper<Object,Text, Text,Intwritable>{
        private final static Intwritable one = new Intwritable(1);
        public void map(Object key,Text value, Context context) throws IOException,
InterruptedException{
        String line = value.toString();
        // 指定逗号为分隔符，组成数组
        String array[] = line.split(", ");
        // 提取数组中的访问日期作为 Key
        String keyoutput = array[1];
        // 组成键值对
        context.write(new Text (keyoutput), one);
```

```
    }
}

public static class MyReducer extends Reducer<Text, Intwritable, Text,Intwritable>{
    private Intwritable result -new Intwritable ();
    public void reduce (Text key, Iterable<Intwritable> values,Context context)
 throws IOException, InterruptedException{
        // 定义累加器，初始值为 0
        int sum = 0;
        for (IntWritable val : values) {
            // 将相同键的所有值进行累加
            sum += val.get ();
        }
        result.set (sum);
        context.write (key, result);
    }
}

public static void main(String[] args) throws Exception{
    Configuration conf = new Configuration();
    Job job=Job.getInstance(conf, "Daily Access Count");
    job.setJarByclass (dailyAccessCount.class);
    job.setMapperClass (MyMapper.class);
    job.setReducerClass(MyReducer.class);
    job.setMapoutputKeyclass(Text.class);
    job.setMapoutputValueClass(IntWritable.class);
    job.setoutputKeyclass (Text.class);
    job.setOutputValueClass (Intwritable.class);
    for (int i=0;i<args.length-1;++i){
        FileInputFormat.addInputPath(job,new Path(args[i]));
    }
    FileoutputFormat.setOutputPath(job,new Path (args[args. length - 1]));
    System.exit(job.waitForCompletion(true) ? 0 : 1);
}
```

2. 需求二编写代码实现按访问次数排序的处理逻辑

① 分析主要模块的处理逻辑，主要是 Mapper 与 Reducer 两个模块。此外，还有输入键值对与输出键值对的定义。

② 编写代码来实现各个核心模块。

③ 补充与完善代码文件 accessTimesSort.java，完整内容如下。

```
package test;
import java.io.IOException;
import org. apache.hadoop.conf.Configuration;
import org. apache.hadoop.fs.Path;
import org. apache.hadoop.io. IntWritable;
import org. apache.hadoop.io.Text;
import org.apache.hadoop.mapreduce.Job;
import org.apache.hadoop.mapreduce.Mapper;
import org. apache.hadoop.mapreduce.Reducer;
```

```
import org.apache.hadoop.mapreduce.lib.input.FileInputFormat;
import org.apache.hadoop.mapreduce.lib.output.FileoutputFormat;
import org.apache.hadoop.util.GenericoptionsParser;

public class accessTimesSort{

    public static class MyMapper extends Mapper<Object,Text,Intwritable,Text>{
        public void map(Object key,Text value, Context context) throws
IOException, InterruptedException{
            String line= value.toString();
            // 指定tab为分隔符，组成数组
            String array[] = line.split("\t ");
            // 提取访问次数作为Key
            int keyoutput = Integer.parseInt(array[1]);
            // 提取访问日期作为value
            String valueOutput = array[0];
            // 组成键值对
            context.write(new Intwritable(keyoutput), new Text(valueOutput));
        }
    }

public static class MyReducerextends extends Reducer<Intwritable, Text, Text,
Intwritable>{
    public void reduce (Intwritable key, Iterable<Text> values,Context context)
 throws IOException, InterruptedException{
            for (Text value : values){
                context.write(value, key);
            }
        }
    }
    public static void main(String[] args) throws Exception{
        Configuration conf = new Configuration ();
        Job job = Job.getInstance (conf, "Access Time Sort");
        job.setJarByClass(accessTimessort.class);
        job.setMapperClass(MyMapper.class);
        job.setReducerClass(MyReducer.class);
        job.setMapoutputKeyClass(IntWritable.class);
        job.setMapoutputValueClass(Text.class);
        job.setOutputKeyClass(Text.class);
        job.setoutputValueclass (Intwritable.class);
        for (int i = 0;i< args.length - 1;++ i) {
            FileInputFormat.addInputPath(job,new Path(args[i]));
        }
        FileoutputFormat.setoutputPath(job,new Path(args[args.length - 1]));
        System.exit(job.waitForCompletion(true) ? 0 :1);
    }
}
```

第二部分，编译生成JAR包并上传到Hadoop集群上执行。

① 编译生成 accessTimesSort.jar。

② 上传 accessTimesSort.jar 到 Hadoop 集群服务器节点。

③ 在 Hadoop 集群服务器的终端以 hadoop jar 命令提交任务，具体命令如下。

```
hadoop jar accessTimessort.jar /user/root/AccessCount /user/root/TimesSort
```

④ 检查输出文件目录的结果，按访问次数排序的结果如图 4-8 所示。

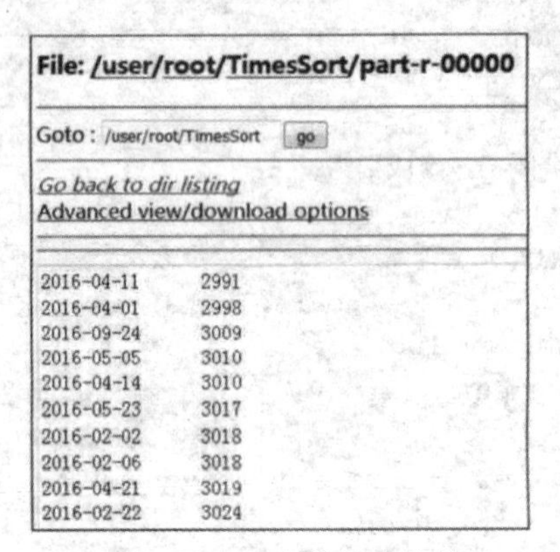

图 4-8　按访问次数排序的结果

图 4-8 中有两列数据：第一列是访问日期；第二列是对应日期的总访问次数，所有记录已按访问次数由小到大进行排序。

4.3 本章小结

本章介绍了 MapReduce 编程的基础知识。MapReduce 把复杂的、运行在 Hadoop 集群上的并行计算过程集成到了两个模块（Mapper 与 Reducer）中。开发人员只要把业务处理逻辑通过其中的 map 函数与 reduce 函数来实现，就可以达到分布式并行编程的目的。

MapReduce 执行过程主要包括以下部分：读取分布式文件系统中的数据，进行数据分片，执行 Map 任务以输出中间结果，Shuffle 阶段对中间结果进行汇合与排序，再传到 Reduce 任务，在 Reduce 阶段对数据进行处理，输出最终结果到分布式文件系统。

第 5 章 Hive 部署与编程基础

Hive 是一个构建于 Hadoop 顶层的数据仓库工具，主要用于对存储在 Hadoop 中的文件数据集进行数据整理、特殊查询和分析处理。它在某种程度上可以看作用户编程接口，其本身不存储和处理数据，而是依赖 HDFS 存储数据，依赖 MapReduce 处理数据。

【学习目标】

1．掌握 Hive 的基础知识。

2．掌握伪分布式 Hive 的搭建方法。

3．熟练运用 Hive 基本操作命令。

4．能对 Hive 进行综合运用。

5.1 搭建伪分布式 Hive

【任务描述】实现 Hive 的伪分布式部署。

Hive 是基于 Hadoop 的一个数据仓库工具，可以将结构化的数据文件映射为一张数据库表，并提供简单的 SQL 查询功能，可以将 SQL 语句转换为 MapReduce 任务进行运行。要掌握 Hive 的伪分布式部署，就要对 Hive 有一个全面的认识。

5.1.1 Hive 概述

1．概念

Hive 可以将结构化的数据文件映射为一张数据库表，并提供类似于关系数据库语言 SQL 的查询工具 HiveQL。我们可以通过 HiveQL 快速实现简单的数据统计。Hive 自身可以将 HiveQL 语句转化为 MapReduce 任务进行运行，而不必开发专门的 MapReduce 应用，这可以降低我们操作大数据的成本，因此，Hive 也十分适合数据仓库的统计分析。

2．优缺点

如果直接使用 MapReduce 对数据进行整理、特殊查询和分析存储，则会面临很多问题，例如人员学习成本太高、复杂查询的开发难度太大等。在这种情况下，我们可以选择使用 Hive 处理数据。

Hive 的优点如下。

① 简单、易上手。Hive 提供了类 SQL 查询语言 HQL，最大限度地实现了和 SQL 标准的兼容，这大大降低了传统数据分析人员的学习难度，提供了快速开发的能力。

② 可扩展。Hive 为超大数据集设计了计算/扩展能力（MapReduce 作为计算引擎，HDFS 作为存储系统），一般情况下不需要重启服务，便可以自由扩展集群的规模。

③ Hive 提供统一的元数据管理（Derby、MySQL 等），并可与 Pig、Presto 等共享。

④ 延展性强。Hive 支持用户自定义函数，用户可以根据自己的需求来实现自己的函数。

⑤ 容错性好。Hive 具有良好的容错性，若节点出现问题，SQL 仍可完成执行。

Hive 的缺点如下。

① Hive 的 HQL 语言表达能力有限，无法表达迭代算法。

② Hive 不擅长数据挖掘，由于受到 MapReduce 数据处理流程的限制，因此无法实现效率更高的算法。

③ Hive 自动生成的 MapReduce 作业通常不够智能化。

④ Hive 调优比较困难，颗粒度较粗。

⑤ Hive 不擅长处理实时性高的场合。

3．数据类型

Hive 支持的数据类型可以分为基础数据类型和复杂数据类型两大类，其中的基础数据类型如表 5-1 所示。

表 5-1　Hive 支持的基础数据类型

数据类型	说明	例子
TINYINT	1B 有符号整数（-128～127）	20
SMALLINT	2B 有符号整数（-32768～32767）	20
INT（INTEGER）	4B 有符号整数（-2147483648～2147483647）	20
BIGINT	8B 有符号整数（-9223372036854775808～9223372036854775807）	20
FLOAT	4B 单精度浮点数	3.14
DOUBLE	8B 双精度浮点数	3.14
DECIMAL	任意精度的带符号小数	Decimal(3, 1)，其中，3 表示数字的长度，1 表示小数点后的位数，取值范围为-99.9～99.9
STRING	字符串，变长	'a'、'b'、'1'
VARCHAR	变长字符串	'a'
CHAR	固定长度字符串	'a'
TIMESTAMP	时间戳，纳秒精度	122327493795
DATE	日期 YYYY-MM-DD（'0000-01-01'～'9999-12-31'）	'2013-01-01'
BOOLEAN	布尔类型	true/false
BINARY	字节数组	

Hive 支持的复杂数据类型如表 5-2 所示，这些复杂类型由基础数据类型组成。

表 5-2 Hive 支持的复杂数据类型

数据类型	描述	字面语法示例
STRUCT	一组命名的字段，字段类型可以不同	Struct ('a', 1, 2.0)
MAP	一组无序的键值对，其中，键的类型必须是原始数据类型，值可以是任何类型。同一个映射的键的类型必须相同，值的类型也必须相同	Map ('a', 1)
ARRAY	有序数组，字段的类型必须相同	Array (1, 2)

5.1.2 Hive 安装和配置

Hive 的安装模式一共有 3 种：内嵌模式、本地模式、远程模式。

（1）内嵌模式

内嵌模式是指 Hive 内嵌在 derby 数据库中来存储元数据，不需要额外启动 metastore 服务。数据库和 metastore 服务都嵌入主 Hive Server 进程中。这种模式是默认的，配置起来比较简单，但是一次只能连接一个客户端，适用于实验，不适用于生产环境。

（2）本地模式

本地模式采用外部数据库来存储元数据，目前支持的数据库有 MySQL、Postgres、Oracle、MS SQL Server。目前这种模式对 MySQL 的支持是最好的，对其他数据库的支持（例如 Oracle）还不是很好。

本地模式不需要单独启动 metastore 服务，用的是和 hive 服务在同一个进程中的 metastore 服务。也就是说，当我们启动一个 hive 服务时，本地模式会默认帮我们启动一个 metastore 服务。

（3）远程模式

远程模式需要单独启动 metastore 服务，每个客户端都在配置文件中配置相关内容，以连接该 metastore 服务。远程模式的 metastore 服务和 hive 服务运行在不同的进程中。

在生产环境中，读者可以根据实际情况选择本地模式或远程模式，下面详细介绍本地模式的安装和配置方法。

1. 下载安装文件

从 Hive 的官方网站下载 Hive 安装文件，本书采用的 Hive 版本是 3.1.2。安装 Hive 之前需要先把 JDK 安装好。读者可根据实际情况选择合适的 Hive 版本。

安装文件下载后，需要进行解压。我们将 Hive 安装在/opt/目录下，使用以下命令对安装文件解压、重命名和权限修改。

```
$ tar -zxvf apache-hive-3.1.2-bin.tar.gz -C /opt
$ cd /opt
$ mv apache-hive-3.1.2-bin hive
$ chown -R 755 hive
```

2. 配置环境变量

为了方便后续使用，我们将 Hive 中可执行命令的路径加入环境变量 PATH，之后便可在任意目录下直接运行 Hive 的命令。使用 vim 编辑器打开～/.bashrc 文件进行编辑，命令如下。

```
$ vim ～/.bashrc
```

在打开文件的末尾添加以下内容。

```
export HIVE_HOME = /opt/hive
export PATH = $PATH:$HIVE_HOME/bin
```

保存～/.bashrc 文件并退出编辑器。为了使上面的配置项立即生效，我们运行以下的命令。

```
$ source ～/.bashrc
```

3．修改配置文件

使用 vim 编辑器对 hive-site.xml 文件进行配置。

```
$ vim hive-site.xml
```

在打开的文件中进行内容修改，具体如下。

```
<?xml version = "1.0" encoding = "UTF-8" standalone = "no"?>
<?xml-stylesheet type = "text/xsl" href = "configuration.xsl"?>
<configuration>
<property>
   <name>javax.jdo.option.ConnectionURL</name>
   <value>jdbc:mysql:// localhost:3306/hive?useSSL = false</value>
</property>
<property>
   <name>javax.jdo.option.ConnectionDriverName</name>
   <value>com.mysql.jdbc.Driver</value>
</property>
<property>
   <name>javax.jdo.option.ConnectionUserName</name>
   <value>hive</value>
</property>
<property>
   <name>javax.jdo.option.ConnectionPassword</name>
   <value>hive</value>
</property>
</configuration>
```

注意：以上内容在配置时需要根据实际情况进行配置，例如，用户名和密码都要如实填写。

4．安装并配置 MySQL

（1）安装 MySQL

安装 MySQL 之前，需要先更新软件源，所用命令如下。

```
$ sudo apt-get update
```

更新完软件源之后，执行以下命令安装 MySQL。

```
$ sudo apt-get install mysql-server
```

在安装过程中，读者需根据提示设置用户名和密码。本书的用户名和密码均为 Hive。设置完成后等待软件自动安装完毕即可。

（2）启动 MySQL 服务

一般数据库安装完后会自动启动 MySQL，这时可以使用以下命令查看 MySQL 是否正常启动。

```
$ sudo netstat-tap | grep mysql
```

如果 MySQL 节点处于 LISTEN 状态，则表示启动成功。手动启动、关闭 MySQL 数据库的命令如下。

```
$ service mysql start
$ service mysql stop
```

（3）下载 MySQL JDBC 驱动

要让 Hive 能够连接到 MySQL 数据库，需要 MySQL JDBC 驱动。我们可以在 MySQL 官网下载 mysql-connector-java-5.1.46.jar.gz（根据实际情况选择合适的版本即可），解压驱动包，并将 mysql-connector-java-5.1.46.jar 文件复制一份到 Hive 的 lib 目录下，命令如下。

```
$ tar -zxvf mysql-connector-java-5.1.46.jar.gz
$ cd mysql-connector-java-5.1.46
$ cp mysql-connector-java-5.1.46.jar /opt/hive/lib
```

（4）创建数据库

在 MySQL 中创建 hive 数据库，用于保存 Hive 的数据。hive 数据库名与 Hive 配置文件 hive-site.xml 中的 jdbc:mysql://localhost:3306/hive 相对应。创建数据库的命令需要在 MySQL Shell 界面的“mysql>”命令提示符下执行，具体如下。

```
mysql> create database hive
```

（5）配置 MySQL 允许 Hive 接入

对 MySQL 进行权限配置，以允许 Hive 接入 MySQL。在 MySQL Shell 界面执行以下命令即可。

```
mysql> grant all on *.* to hive@localhost identified by 'hive';
mysql> flush privileges;
```

第一个命令表示将 MySQL 的所有数据库表的所有权限赋给 hive 用户，第二个命令表示刷新 MySQL 系统权限关系表。

（6）启动 Hive

Hive 是基于 Hadoop 的数据仓库，因此，在启动 Hive 之前，需要先启动 Hadoop 集群。启动命令如下。

```
$ start-all.sh
```

以上启动方式是基于已将 Hadoop 可执行命令的路径配置到 PATH 变量中这种情况。如果没有配置路径，则需要先将目录切换到 Hadoop 安装目录的 sbin 目录下执行上面的启动命令，然后启动 Hive。启动命令如下。

```
$ hive
```

和启动 Hadoop 情况类似，Hive 启动的前提是已将 Hive 可执行命令的路径配置到 PATH 变量中。如果没有配置路径，则需要先将目录切换到 Hive 安装目录的 bin 目录下，再执行上面的启动命令。

5.2 Hive 基本操作

【任务描述】熟练使用 HiveQL 对数据库进行增、删、改、查等操作。

HiveQL 是 Hive 的查询语言。它和 SQL 语言类似，可以通过编写 HiveQL 语句实现对数据库的操作。

5.2.1 数据库基本操作

1．创建数据库

（1）语法

```
CREATE DATABASE [IF NOT EXISTS] database_name
[COMMENT database_comment]
[LOCATION hdfs_path]
[WITH DBPROPERTIES (property_name = property_value, ...)];
```

（2）案例

创建一个数据库，命令如下。

```
hive> create database db_hive1;
```

判断要创建的数据库是否存在，若不存在，则创建数据库，命令如下。

```
hive> create database if not exists db_hive2;
```

创建一个数据库，带有 dbproperties 参数，命令如下。

```
hive> create database db_hive3 with dbproperties('create_date' = '2023-07-31');
```

2．查看数据库信息

（1）语法

```
SHOW DATABASES [LIKE 'identifier_with_wildcards'];
```

（2）案例

```
hive> show databases like 'db_hive*';
OK
db_hive1
db_hive2
```

3．修改数据库

ALTER DATABASE 命令可以修改数据库的某些信息，其中包括 DBPROPERTIES、LOCATION、OWNER USER。需要注意的是，修改数据库 LOCATION 的操作不会改变当前已有表的路径信息，只是改变后续创建的新表的默认父目录。

（1）语法

```
--修改 DBPROPERTIES
ALTER DATABASE database_name SET DBPROPERTIES
                                       (property_name = property_value, ...);

--修改 LOCATION
ALTER DATABASE database_name SET LOCATION hdfs_path;

--修改 OWNER USER
ALTER DATABASE database_name SET OWNER USER user_name;
```

（2）案例

修改 DBPROPERTIES，命令如下。

```
hive> alter database db_hive3 set dbproperties ('create_date' = '2023-08-01');
```

4．删除数据库

（1）语法

```
DROP DATABASE [IF EXISTS] database_name [RESTRICT|CASCADE];
```

注：RESTRICT 为严格模式，若数据库不为空，则删除失败，该模式默认为模式；CASCADE 为级联模式，若数据库不为空，则数据库中的表也会被一起删除。

（2）案例

删除空数据库，命令如下。

```
hive> drop database db_hive2;
```

删除非空数据库，命令如下。

```
hive> drop database db_hive3 cascade;
```

5.2.2　数据表基本操作

1．创建数据表

（1）完整语法

```
CREATE [TEMPORARY] [EXTERNAL] TABLE [IF NOT EXISTS] [db_name.]table_name
[(col_name data_type [COMMENT col_comment], ...)]
[COMMENT table_comment]
[PARTITIONED BY (col_name data_type [COMMENT col_comment], ...)]
[CLUSTERED BY (col_name, col_name, ...)
[SORTED BY (col_name [ASC|DESC], ...)] INTO num_buckets BUCKETS]
[ROW FORMAT row_format]
[STORED AS file_format]
[LOCATION hdfs_path]
[TBLPROPERTIES (property_name = property_value, ...)];
```

（2）关键字说明

EXTERNAL：外部表，与之对应的是内部表（管理表）。内部表（管理表）意味着 Hive 会完全管理该表，其中包括元数据和 HDFS 中的数据。外部表则意味着 Hive 只接管元数据，不管理 HDFS 中的数据。

PARTITIONED BY：创建分区表。

ROW FORMAT：指定 SERDE，SERDE 是 Serializer and Deserializer（序列化和反序列化）的简写。Hive 使用 SERDE 对每行数据进行序列化和反序列化处理。

DELIMITED：表示对文件中的每个字段按照特定分割符进行分割，使用默认的 SERDE 对每行数据进行序列化和反序列化处理。

fields terminated by：列分隔符。

LOCATION：指定表所对应的 HDFS 路径，若不指定路径，其默认值为${hive.metastore.warehouse.dir}/db_name.db/table_name。

（3）案例

Hive 中默认创建的表都是内部表。创建内部表的命令如下。

```
hive> create table if not exists student(
```

```
    id int,
    name string
)
```

外部表通常用于处理其他工具上传的数据文件。创建外部表的命令如下。

```
hive> create external table if not exists student(
    id int,
    name string
)
```

创建 student 表，该表包含两个属性——id、name，可以读取路径/user/hive/warehouse/student 下以“\t”分割的数据。命令如下。

```
hive> create external table if not exists student(
    id int,
    name string
)
row format delimited fields terminated by '\t'
location '/user/hive/warehouse/student';
```

Hive 中的分区是把一张大表的数据按照业务需要分散地存储到多个目录下，每个目录是该表的一个分区。查询时可通过 where 语句中的表达式选择所需要的分区，这样查询效率会提高很多。

创建分区表的命令如下。

```
hive> create table dept_partition
(
    deptno int,     --部门编号
    dname  string, --部门名称
    loc    string  --部门位置
)
    partitioned by (day string)
row format delimited fields terminated by '\t';
```

2．查看数据表信息

（1）展示所有表

① 语法

```
SHOW TABLES [IN database_name] LIKE ['identifier_with_wildcards'];
```

② 案例

```
hive> show tables like 'stu*';
```

（2）查看表信息

① 语法

```
DESCRIBE [EXTENDED | FORMATTED] [db_name.]table_name;
```

② 案例

```
hive> describe stu;
```

3．修改数据表

（1）重命名表

① 语法

```
ALTER TABLE table_name RENAME TO new_table_name;
```

② 案例

```
hive> alter table stu rename to stu1;
```

（2）修改列信息

① 语法

增加列语句允许用户增加新的列，新增列的位置位于末尾，命令如下。

```
ALTER TABLE table_name ADD COLUMNS (col_name data_type [COMMENT col_comment], ...);
```

更新列语句允许用户修改指定列的列名、数据类型、注释信息及在表中的位置，命令如下。

```
ALTER TABLE table_name CHANGE [COLUMN] col_old_name col_new_name
column_type [COMMENT col_comment] [FIRST|AFTER column_name];
```

替换列语句允许用户用新的列集替换表中原有的全部列，命令如下。

```
ALTER TABLE table_name REPLACE COLUMNS (col_name data_type [COMMENT
col_comment], ...);
```

② 案例

添加列，命令如下。

```
hive> alter table stu add columns(age int);
```

更新列，命令如下。

```
hive> alter table stu change column age ages double;
```

替换列，命令如下。

```
hive> alter table stu replace columns(id int, name string);
```

4．删除数据表

（1）语法

```
DROP TABLE [IF EXISTS] table_name;
```

（2）案例

```
hive> drop table stu;
```

5.2.3 数据基本操作

前面介绍了数据库和数据表的创建、修改等内容，下面介绍如何操作数据表中的数据，数据表中的数据可以进行加载、查询、插入等操作。

1．数据加载

LOAD 语句可将文件导入 Hive 表中。

（1）语法

```
LOAD DATA [LOCAL] INPATH 'filepath' [OVERWRITE] INTO TABLE tablename
[PARTITION (partcol1 = val1, partcol2 = val2 ,...)];
```

关键字说明如下。

LOCAL：表示从本地加载数据到 Hive 表，否则从 HDFS 加载数据到 Hive 表。

OVERWRITE：表示覆盖表中已有的数据，否则表示追加。

PARTITION：表示上传到指定分区，若目标是分区表，需指定分区。

（2）案例

加载本地文件到 Hive 表中，命令如下。

```
hive> load data local inpath '/opt/module/datas/student.txt' into table student;
```

加载 HDFS 文件到 Hive 表中，命令如下。

```
hive> load data inpath '/user/atguigu/student.txt' into table student;
```

加载数据并覆盖表中已有的数据，命令如下。

```
hive> load data inpath '/user/atguigu/student.txt' overwrite into table student;
```

2. 数据查询

Hive 表数据的查询和 SQL 语句完全一样，我们在这里进行简单介绍。

查询所有的学生信息，命令如下。

```
hive> select * from student;
```

查询学生信息——姓名、年龄，命令如下。

```
hive> select name, age from student;
```

条件查询，命令如下。

```
hive> select name, age from student where age>10;
```

3. 数据插入

向表中批量插入数据的常用方法有两种，一种是前面介绍的表数据装载，另一种是利用子查询将结果集批量插入表中。

（1）语法

```
INSERT (INTO | OVERWRITE) TABLE tablename [PARTITION (partcol1 = val1,
partcol2 = val2 ,...)] select_statement;
```

关键字说明如下。

INTO：将结果追加到目标数据表中。

OVERWRITE：用结果覆盖原有数据。

（2）案例

根据查询结果插入数据并覆盖原有数据，命令如下。

```
hive> insert overwrite table student3
select
   id,
   name
from student;
t * from student where age = 10;
```

5.3 编程实现——部门工资统计

现在我们通过统计各部门工资总和的实例介绍 Hive 的具体使用方法。

1. 数据准备

创建输入数据文件，首先，创建数据文件存储目录，例如/usr/local/hive/input，命令如下。

```
$ mkdir -p /usr/local/hive/input
```

在 input 文件夹中分别创建员工和部门信息文件 emp.csv 和 dept.csv，命令如下。

```
$ touch /usr/local/hive/input/emp.csv
$ touch /usr/local/hive/input/dept.csv
```

利用 vim 编辑器将员工信息和部门信息分别写入 emp.csv 和 dept.csv 文件，其中的员工信息如下（注意字段之间的分隔符）。

```
7369,SMITH,CLERK,7902,1980/12/17,800,,20
7499,ALLEN,SALESMAN,7698,1981/2/20,1600,300,30
7521,WARD,SALESMAN,7698,1981/2/22,1250,500,30
7566,JONES,MANAGER,7839,1981/4/2,2975,,20
7654,MARTIN,SALESMAN,7698,1981/9/28,1250,1400,30
7698,BLAKE,MANAGER,7839,1981/5/1,2850,,30
7782,CLARK,MANAGER,7839,1981/6/9,2450,,10
7788,SCOTT,ANALYST,7566,1987/4/19,3000,,20
7839,KING,PRESIDENT,,1981/11/17,5000,,10
7844,TURNER,SALESMAN,7698,1981/9/8,1500,0,30
7876,ADAMS,CLERK,7788,1987/5/23,1100,,20
7900,JAMES,CLERK,7698,1981/12/3,950,,30
7902,FORD,ANALYST,7566,1981/12/3,3000,,20
7934,MILLER,CLERK,7782,1982/1/23,1300,,10
```

部门信息如下。

```
10,ACCOUNTING,NEW YORK
20,RESEARCH,DALLAS
30,SALES,CHICAGO
40,OPERATIONS,BOSTON
```

2．创建员工/部门信息表

进入 Hive 命令行界面，创建员工信息表，命令如下。

```
hive> create table emp(empno int,ename string,job string,mgr int,hiredate
string,sal int,comm int,deptno int) row format delimited fields terminated by ',';
```

创建部门信息表，命令如下。

```
hive> create table dept(deptno int,dname string,loc string) row format delimited
fields terminated by ',';
```

3．导入数据

将 emp.csv 和 dept.csv 文件中的数据分别加载到员工信息表和部门信息表中，命令如下。

```
hive> load data local inpath '/usr/local/hive/input/emp.csv' into table emp;
hive> load data local inpath '/usr/local/hive/input/dept.csv' into table dept;
```

4．工资统计

编写 HiveQL 语句，实现统计各部门工资总和，具体如下。

```
hive> select dept.dname,sum(emp.sal) as sal from emp,
dept where emp.deptno = dept.deptno group by dept.dname;
```

得到的部门工资统计结果如图 5-1 所示。

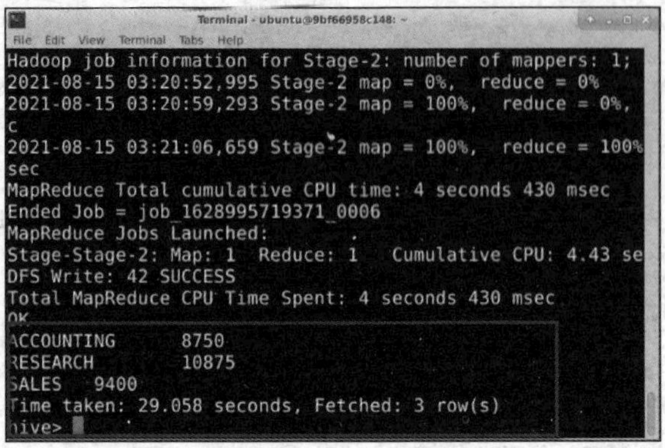
Terminal - ubuntu@9bf66958c148: ~
File Edit View Terminal Tabs Help
Hadoop job information for Stage-2: number of mappers: 1;
2021-08-15 03:20:52,995 Stage-2 map = 0%, reduce = 0%
2021-08-15 03:20:59,293 Stage-2 map = 100%, reduce = 0%,
c
2021-08-15 03:21:06,659 Stage-2 map = 100%, reduce = 100%
sec
MapReduce Total cumulative CPU time: 4 seconds 430 msec
Ended Job = job_1628995719371_0006
MapReduce Jobs Launched:
Stage-Stage-2: Map: 1 Reduce: 1 Cumulative CPU: 4.43 se
DFS Write: 42 SUCCESS
Total MapReduce CPU Time Spent: 4 seconds 430 msec
OK
ACCOUNTING 8750
RESEARCH 10875
SALES 9400
Time taken: 29.058 seconds, Fetched: 3 row(s)
ive>

图 5-1 得到的部门工资统计结果

5.4 本章小结

本章介绍了 Hive 的基础知识、安装和配置方法；还介绍了 Hive 的常用命令。本章通过一个部门工资统计的实例，帮助读者更好地掌握 Hive 的使用方法。

第 6 章 Spark 部署与编程基础

Spark 是 2009 年由马泰·扎哈里亚在美国加利福尼亚大学伯克利分校的 AMPLab 实验室开发的 Hadoop 子项目，经过开源后捐赠给 Apache 软件基金会，成为现在众所周知的 Apache Spark。它是由 Scala 语言实现的专门为大规模数据处理而设计的通用计算引擎，经过多年的发展，现已形成一个高速发展且应用广泛的生态系统。

【学习目标】

1. 了解 Spark 运行原理。
2. 掌握 Scala 语言的用法。
3. 掌握 Python 版本的 Spark 安装方法。

6.1 Spark 的运行原理

【任务描述】理解 Spark 的集群架构及作业的运行流程，了解 Spark 的核心数据集 RDD。

6.1.1 集群架构

Spark 的集群架构如图 6-1 所示，架构中的组件说明，具体如下。

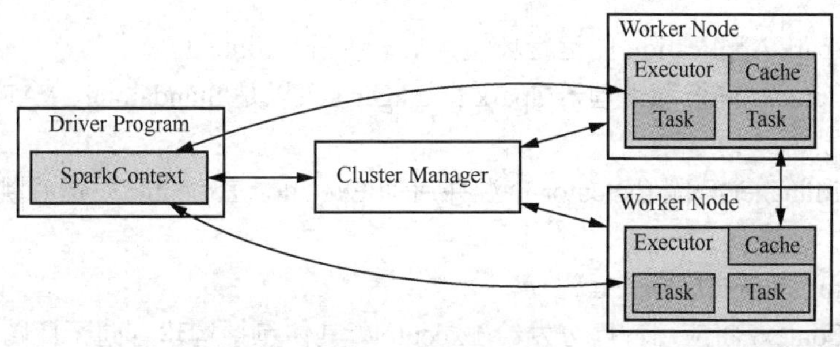

图 6-1 Spark 的集群架构

① Driver Program：运行 Application 的 main 函数并创建 SparkContext。Application 指的是用户编写的 Spark 应用程序，其中包含一个 Driver（驱动器）功能的代码和分布在

集群中多个节点上的 Executor（执行器）代码。

② SparkContext：是所有 Spark 功能的入口，相当于 main()函数。

③ Cluster Manager：资源管理器，指的是在集群上获取资源的外部服务，目前主要有 Standalone 和 Hadoop YARN。Standalone 是 Spark 原生的资源管理器，由主节点负责资源的分配。若使用 YARN，则由 ResourceManager 负责资源的分配。

④ Worker Node：指集群中任何可以运行 Application 的节点，节点上可运行一个或多个 Executor 进程。

⑤ Executor：指运行在 Worker 的 Task（任务）执行器。Executor 启动线程池运行 Task，并且负责将数据存储在内存或磁盘上。每个 Application 会申请 Executor 来处理 Task。

⑥ Task：被送到 Executor 进行处理的具体工作任务。

6.1.2 运行流程

Spark 作业的运行流程如图 6-2 所示。

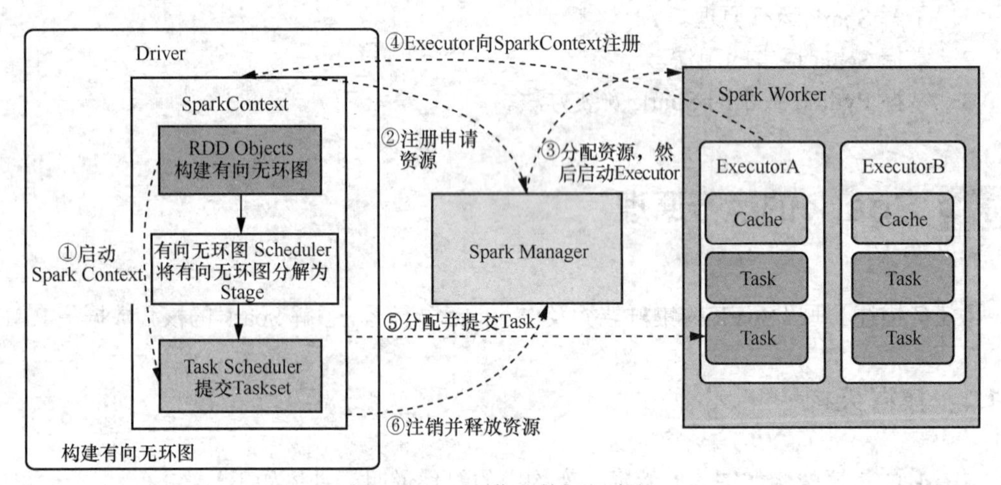

图 6-2 Spark 作业的运行流程

① 构建 Spark Application 的运行环境，启动 SparkContext。

② SparkContext 向资源管理器 Spark Manager（可以是 Standalone、YARN）注册申请运行 Executor 资源。

③ Spark Manager 分配 Executor 资源并启动 Executor。Executor 发送心跳信号给 Spark Manager。

④ Executor 向 SparkContext 注册。

⑤ SparkContext 将应用程序分发给 Executor，具体包括构建有向无环图、将有向无环图分解成 Stage、将 Taskset 发送给 Task Scheduler，以及由 Task Scheduler 将 Task 分配并提交给 Executor。

⑥ Task 在 Executor 上运行，运行完后注销并释放所有资源。

6.1.3　核心数据集 RDD

RDD 是一种可扩展的弹性分布式数据集，是 Spark 最基本的数据抽象，表示一个只读、分区且不变的数据集合，是一种分布式的内存抽象，不具备 Schema 的数据结构，可以基于任何数据结构创建，例如元组、字典和列表等。

RDD 具有对数据进行分片/区、自定义分区计算函数、RDD 之间相互依赖、控制分片数量、使用列表方式进行块存储等特点。

① 分片/区。片/区是 RDD 的基本组成单位。Spark 集群读取一个文件会根据具体的配置将文件加载在不同节点的内存中。每个节点中的数据就是一个分片。对于 RDD 来说，每个分片会被一个计算任务处理，并决定并行计算的颗粒度。用户可以在创建 RDD 时指定 RDD 的分片个数，默认分片个数与 CPU 核心个数相同。

② 自定义分片计算函数。Spark 中 RDD 的计算是以分片为单位的，每个 RDD 都可达到分片计算的目的。

③ RDD 之间相互依赖。RDD 的每次转换都会生成一个新的 RDD，因此 RDD 之间会形成类似于流水线的前后依赖关系。在部分分片数据丢失时，Spark 可以通过这个依赖关系重新计算丢失的分片数据，而不是对 RDD 的所有分片进行重新计算。

④ 控制分片数量。当前 Spark 中实现了两种类型的分片函数，一个是基于哈希的 HashPartitioner，另一个是基于范围的 RangePartitioner。Partitioner 函数不仅决定了 RDD 本身的分片数量，也决定了父 RDD Shuffle 输出时的分片数量。

⑤ 使用列表方式进行块存储。对于 HDFS 来说，列表保存的是每个片/区所在的块的位置。按照“移动数据不如移动计算”的理念，Spark 在进行任务调度的时候，会尽可能地将计算任务分配到它所要处理数据块的存储位置。

6.1.4　核心原理

Spark 核心原理即 Spark 的分布式计算流程，如图 6-3 所示。Spark 的分布式计算流程包含以下步骤。

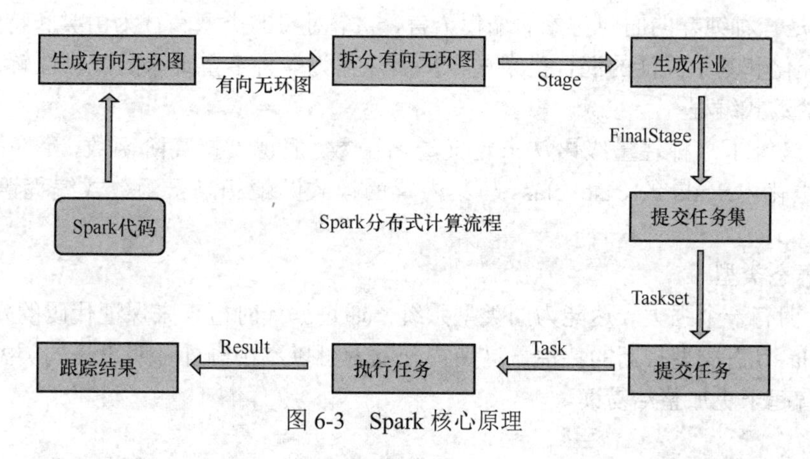

图 6-3　Spark 核心原理

步骤 1：从代码构建有向无环图。

步骤 2：将有向无环图拆分为 Stage。

步骤 3：Stage 生成作业。

步骤 4：FinalStage 提交任务集。

步骤 5：Taskset 提交任务。

步骤 6：Task 执行任务。

步骤 7：Result 跟踪结果。

6.2 Scala 的安装与使用

【任务描述】 Scala 是 Spark 编程常用的语言之一，同时 Spark 是采用 Scala 语言编写的，因此在学习 Spark 之前，需要先了解 Scala 语言的基本概念和 Scala 的安装方法。

6.2.1 Scala 语言概述

Scala 是一种多范式的编程语言，由瑞士洛桑联邦理工学院的马丁・奥德斯于 2001 年创建，其设计初衷是要集成面向对象编程和函数式编程的各种特性，因此，Scala 是一种纯面向对象的语言，每个值都是对象。同时 Scala 也是一种函数式语言，函数也能当成值来使用。由于整合了面向对象编程和函数式编程的特性，Scala 比 Java、C#、C++等语言更加简洁。

Scala 源代码被编译成 Java 字节码，因此它可以运行在 JVM 上，而且可以调用现有的 Java 类库。

6.2.2 Scala 特性

Scala 有四大特性：面向对象、函数式编程、静态类型、可扩展。

（1）面向对象

Scala 是一种纯粹的面向对象的编程语言，一个对象的类型和行为由类和特征来描述。类通过子类化和基于灵活的混合类来进行扩展，作为一种多重继承的可靠性解决方案。

（2）函数式编程

Scala 提供了一种轻量级语法来定义匿名函数，但也支持高阶函数，允许函数嵌套，并支持柯里化。Scala 的 case class 与其内置的模式匹配相当于函数式编程语言中的代数类型。

（3）静态类型

Scala 拥有一个强大表达能力的类型系统，通过编译时检查来保证代码的安全性和一致性。Scala 具有类型推断的特性，这使得开发者可以不用额外标明重复的类型信息，从而让代码看起来更加整洁易读。

（4）可扩展

在实践中，专用领域的应用程序开发往往需要该领域特定的语言扩展。Scala 提供了许多独特语言机制，可以以库的形式轻易、无缝添加新的语言结构。

6.2.3 环境设置与安装

1. 在网页上运行 Scala

在安装 Scala 之前，可以先使用 Scala 在线编程环境来体验。在在线编程环境中，读者可以简单地测试 Scala 代码。下面以打印“Hello world”为例，介绍如何使用在线编程环境测试 Scala 代码。Scala 代码在线测试如图 6-4 所示。具体如下。

首先，通过浏览器访问 ScalaFiddle 官网。

然后，在窗格中输入“println("Hello world")”。

最后，单击“Run”按钮，输出显示在右窗格中。

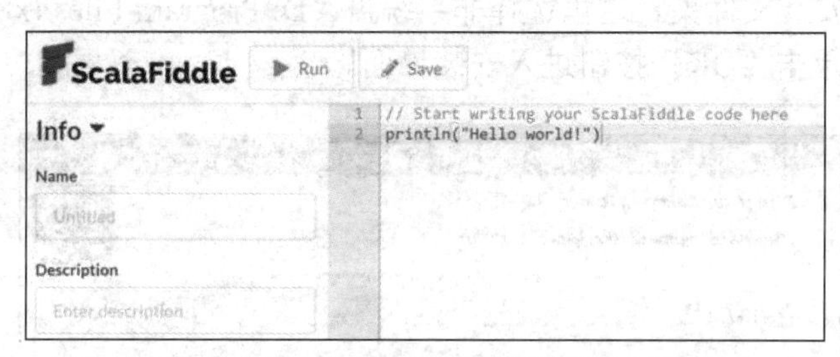

图 6-4 Scala 代码在线测试

2. Scala 环境设置

Scala 的运行环境广泛，可以运行在 Windows、Linux、macOS 等操作系统上。Scala 也是运行在 JVM 上的语言，所以环境中也必须安装 Java，而且 Java 版本必须与安装的 Spark 的 JDK 编译版本一致。本书使用的是 JDK 8（Java 1.8）。用户自行查看本地 Java 版本并配置好环境变量。

我们通过 java –version 命令查看 Java 版本，如图 6-5 所示。

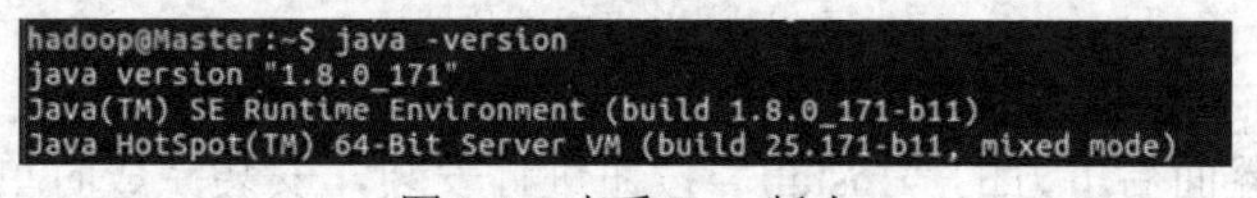

图 6-5 查看 Java 版本

3. Scala 安装

（1）在 Linux 和 macOS 操作系统上安装 Scala

首先从 Scala 官网下载压缩文件包 scala-2.12.8.tgz 并上传到/opt 目录，使用以下命令解压文件到/usr/local 目录下。

```
$sudo tar -zxf scala-2.12.8.tgz -C /usr/local/
```

为了更方便地使用 Scala 进行编程，我们使用 sudo vim /etc/profile 命令打开配置文件

/etc/profile，设置 Scala 环境变量。配置 Scala 开发环境如图 6-6 所示，在配置文件末尾处输入 Scala 解压后的地址，并添加在 PATH 环境变量，之后退出编辑模式并保存文件。我们使用 source /etc/profile 命令重新加载/etc/profile 配置文件。

```
export SCALA_HOME=/usr/local/scala-2.11.8
export PATH=$SCALA_HOME/bin:$PATH
```

图 6-6　配置 Scala 开发环境

（2）在 Windows 操作系统下安装 Scala

在 Windows 操作系统下安装 Scala 的步骤如下。

步骤 1：从 Scala 官网下载 scala.msi 文件。

步骤 2：双击 scala.msi 文件，开始安装软件。

步骤 3：进入欢迎界面，单击右下方的“Next”按钮后出现许可协议选择提示框，选择接受许可协议中的条款并单击右下方的“Next”按钮。

步骤 4：选择安装文件夹，这里选择非系统盘“D:\Program Files (x86)\Spark\scala”，如图 6-7 所示，并单击“OK”按钮进入安装界面。

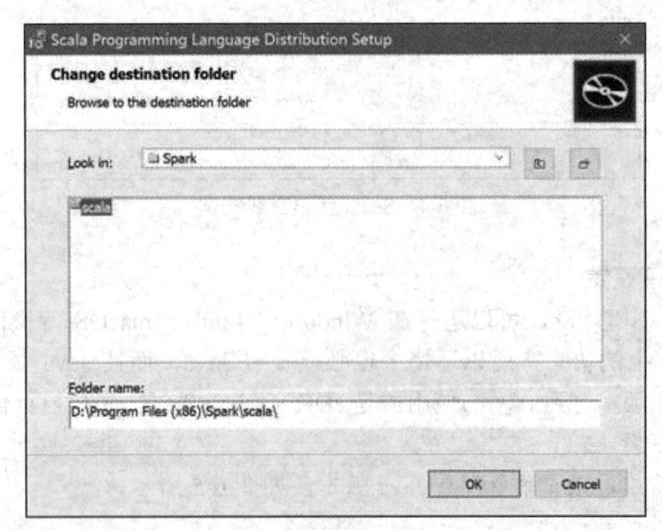

图 6-7　选择安装位置

① 单击安装界面右下方的“Install”按钮进行安装，安装完成时单击“Finish”按钮完成安装。

② 配置环境变量，依次单击“计算机”→“属性”→“高级系统设置”→“环境变量”，选择“path”变量，在该变量后面添加 Scala 安装包\bin 目录所在的路径，例如“D:\Program Files (x86)\Spark\scala\bin”。

4．运行 Scala

Scala 解释器的交互式模式也称为读取-计算-打印-循环（Read-evaluate-print Loop，REPL）。当在命令行中输入 scala 命令时，就会进入 REPL。启动 Scala 后会看到图 6-8 所示的信息。

```
hadoop@Master:~$ scala
Welcome to Scala 2.12.8 (Java HotSpot(TM) 64-Bit Server VM, Java 1.8.0_171).
Type in expressions for evaluation. Or try :help.

scala>
```

图 6-8　启动 Scala 后的信息

REPL 可提供立即产生交互反馈的体验，这对于读者来说非常有用。在图 6-9 中，我们先执行基本的算术运算，然后输入:quit 命令退出 REPL。

```
scala> 52*77
res0: Int = 4004

scala> :quit
hadoop@Master:~$
```

图 6-9　执行算术运算并退出 REPL

REPL 提供了粘贴模式，用于粘贴大量的代码块。在 REPL 中输入:paste 命令，进入粘贴模式，即可粘贴大量的代码块。之后，通过组合键 Ctrl+D 可退出粘贴模式。REPL 的粘贴模式示例如图 6-10 所示。

图 6-10 所示代码是一个类，该类包含一个实现两个数相加的方法。如果要使用该方法，可通过 import 命令进行加载。加载方法如图 6-11 所示，其中，add 表示类名，addInt 表示方法名。

```
scala> :paste
// Entering paste mode (ctrl-D to finish)

object add{
    def addInt(a:Int, b:Int):Int={
        var sum:Int=0
        sum=a+b
        return sum
    }
}

// Exiting paste mode, now interpreting.

defined object add
```

图 6-10　REPL 的粘贴模式示例

```
scala> import add.addInt;
import add.addInt

scala> addInt(2,3);
res0: Int = 5
```

图 6-11　加载方法

注意：Scala 语句末尾的分号是可选的。若一行中仅有一个语句，则可不加分号；若一行中包含多条语句，则需要使用分号把不同的语句分隔开。

6.3 Spark 的安装与使用

【任务描述】掌握 Spark 环境的搭建方法和 Spark 服务监控的查看方法。

6.3.1　环境搭建前的准备

Spark 使用 Scala 语言进行开发，Scala 运行在 Java 平台上，因此需要下载并安装 JDK 和 Scala。值得注意的是，Scala、Java 和 Spark 三者之间是有版本搭配限制的，读者可以根据官方文档提供的组合进行下载。Spark、Java、Scala 具体的版本对应关系可在官网相关文档中看到，（部分）如图 6-12 所示，本书使用的环境组合是 Spark 2.3.0+Java 8+Scala 2.11。

Downloading

Get Spark from the downloads page of the project website. This documentation is for Spark version 2.3.0. Spark uses Hadoop's client libraries for HDFS and YARN. Downloads are pre-packaged for a handful of popular Hadoop versions. Users can also download a "Hadoop free" binary and run Spark with any Hadoop version by augmenting Spark's classpath. Scala and Java users can include Spark in their projects using its Maven coordinates and in the future Python users can also install Spark from PyPI.

If you'd like to build Spark from source, visit Building Spark.

Spark runs on both Windows and UNIX-like systems (e.g. Linux, Mac OS). It's easy to run locally on one machine — all you need is to have java installed on your system PATH, or the JAVA_HOME environment variable pointing to a Java installation.

Spark runs on Java 8+, Python 2.7+/3.4+ and R 3.1+. For the Scala API, Spark 2.3.0 uses Scala 2.11. You will need to use a compatible Scala version (2.11.x).

Note that support for Java 7, Python 2.6 and old Hadoop versions before 2.6.5 were removed as of Spark 2.2.0. Support for Scala 2.10 was removed as of 2.3.0.

图 6-12　Spark、Java、Scala 的版本对应关系（部分）

Spark 运行在 Linux 环境下，因此在环境搭建之前先需要部署 Linux 环境。这里的 Linux 环境可以是物理机，也可以是虚拟机。具体的部署方法此处不讲解，本书采用 Linux 操作系统的发行版 Ubuntu 作为演示系统。

（1）下载 Spark 安装包

打开浏览器进入 Spark 官网，下载安装包 spark-3.3.0-bin-hadoop3.tgz。

（2）下载 Scala 安装包

打开浏览器进入 Scala 官网，下载安装包 scala-2.12.8.tgz。

（3）下载 java 安装包

打开浏览器进入 JDK 下载页面，下载安装包 jdk-8u171-linux-x64.tar.gz。

6.3.2　Spark 的安装与配置

在使用 Spark 之前，需要先进行相关配置，主要包括安装 SSH、实现免密登录、修改环境变量、修改 Spark 文件夹的访问权限、节点参数配置等。

1．进入 Ubuntu 操作系统

打开 Terminal 命令窗口，如图 6-13 所示。

图 6-13　Terminal 命令窗口

2．关闭防火墙

① 查看防火墙状态，命令如下。

```
$sudo ufw status
Status: active// 防火墙处于启用状态
```

② 如果防火墙处于运行状态，则关闭防火墙，命令如下。

```
$sudo ufw disable
```

系统在启动时会自动禁用防火墙。

③ 查看防火墙状态，命令如下。

```
$sudo ufw status
Status: inactive// 防火墙处于关闭状态
```

3．确认 SSH 是否已安装

① 查看 openssh-client、openssh-server 是否已安装，命令如下。

```
$dpkg  -l | grep  openssh
```

② 如果没有安装 SSH，则进行安装，命令如下。

```
$sudo  apt-get  install  openssh-client
$sudo  apt-get  install  openssh-server
```

4．设置免密登录

① 通过 ssh-keygen 生成一个 RSA 的密钥对，命令如下（按回车键结束）。

```
$ssh-keygen  -t  rsa  -P  ''
```

② 进入目录.ssh，命令如下。

```
$cd  ~/.ssh/
```

③ 生成密钥对，并一直按回车键结束，命令如下（当遇到 y/n 时选 y）。

```
$ssh-keygen  -t  rsa
```

④ 执行以下命令生成授权文件。

```
$cat  id_rsa.pub>>authorized_keys
```

⑤ 将公钥追加到~/.ssh/authorized_keys 文件中，命令如下。

```
$cp ~/.ssh/id_rsa.pub ~/.ssh/authorized_keys// 这很重要
```

⑥ 修改映射，命令如下。

```
$sudo  vim  /etc/hosts
```

⑦ 添加映射命令如下。得到的结果如图 6-14 所示。

```
127.0.0.1    Master
```

```
127.0.0.1        Localhost
127.0.0.1        Master
```

图 6-14　添加映射的结果

⑧ 测试免密登录，命令如下。

```
$ssh localhost
```

5．安装 JDK

① 将安装包解压到指定目录，命令如下。

```
$sudo  tar  zxvf  jdk-8u171-linux-x64.tar.gz  -C  /usr/local/
```

② 配置环境变量，命令如下。

```
$sudo  vim  /etc/profile
```

③ 在文件末尾插入环境变量，命令如下。

```
export JAVA_HOME=/usr/local/jdk1.8.0_171
export PATH=$JAVA_HOME/bin:$PATH
export CLASSPATH=$JAVA_HOME/lib/dt.jar:$JAVA_HOME/lib/tools.jar:.
```

④ 使变量生效，命令如下。

```
$source  /etc/profile
```

6．安装 Scala

① 将安装包解压到指定目录，命令如下。

```
$sudo  tar  zxvf  scala-2.12.8.tgz  -C  /usr/local/
$ln  -s  scala-2.12.8  scala
```

② 配置环境变量，命令如下。

```
$sudo vim /etc/profile
```

③ 在文件末尾插入环境变量，命令如下。

```
export SCALA_HOME=/usr/local/scala-2.12.8
export PATH=$SCALA_HOME/bin:$PATH
```

④ 使变量生效，命令如下。

```
$source  /etc/profile
```

7．搭建 Spark 的伪分布

① 将安装包进行解压，并配置环境变量，命令如下。

```
$sudo  tar  zxvf  spark-3.3.0-bin-hadoop3.tgz  -C /usr/local/
```

② 重命名 spark-3.3.0-bin-hadoop3 文件夹为 spark，命令如下。

```
$mv  spark-3.3.0-bin-hadoop3  spark
```

③ 创建超链接（便于使用），命令如下。

```
$ln  -s  /usr/local/spark  /home/hadoop/spark
```

④ 增加环境变量（如果已经安装过 hadoop，此步可以省略，以避免冲突），命令如下。

```
$sudo  vim  /etc/profile
```

⑤ 在文件末尾插入环境变量，命令如下。

```
export SPARK_HOME=/usr/local/spark
export PATH=$SPARK_HOME/bin:$SPARK_HOME/sbin:$PATH
```

⑥ 使变量生效，命令如下。

```
$source  /etc/profile
```

8．修改配置文件

① 进入 Spark 配置文件所在目录，修改 spark-env.sh 文件，命令如下。

```
$cd /usr/local/spark/conf
$cp spark-env.sh.template spark-env.sh
$vi spark-env.sh
```

② 在文件末尾添加以下内容。

```
export JAVA_HOME = /usr/local/jdk1.8.0_171
export SPARK_MASTER_HOST = Master
export SPARK_MASTER_PORT = 7077
```

```
SPARK_LOCAL_IP = localhost
```

③ 修改 slaves 文件，命令如下。

```
$cp slaves.template slaves
$vi slaves
```

④ 将文件中的 localhost 改为主机名 Master，命令如下。

```
# A Spark Worker will be started on each of the machines listed below.
# localhost
Master
```

⑤ 启动 Spark，命令如下。

```
$cd  /usr/local/spark
$sbin/start-all.sh
```

⑥ 进行验证。首先查看进程，命令如下。

```
$jps
```

得到以下进程。

```
1398 Worker
1327 Master
```

然后打开网页 http://localhost:8080（localhost 表示 IP 地址），验证 Spark 是否启动成功。得到的结果如图 6-15 所示。

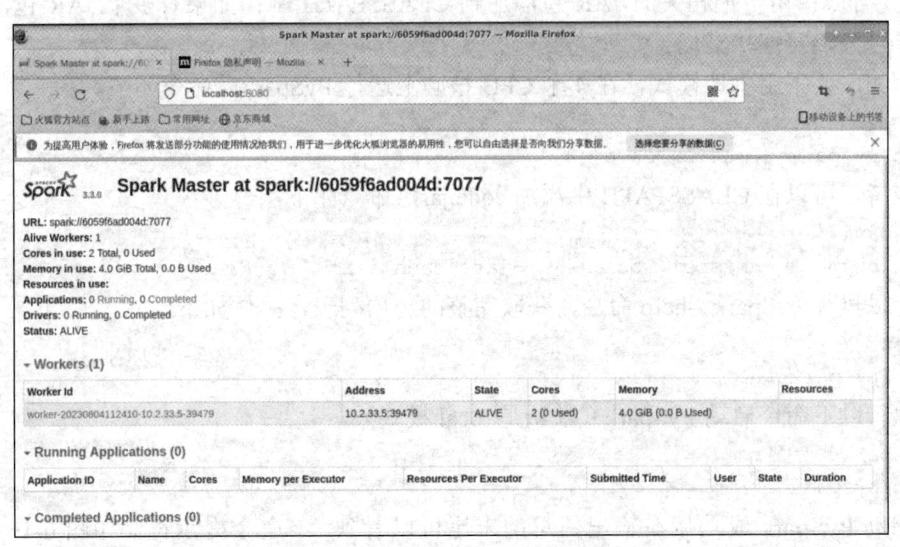

图 6-15　网页验证启动 Spark 的结果

6.3.3　在 PySpark 中运行代码

PySpark 提供了简单的方式来帮助读者学习 Spark API，让读者可以以实时、交互的方式分析数据；同时还提供了 Python 交互式执行环境。

PySpark 命令及其常用的参数如下。

```
pyspark --master <master-url>
```

Spark 的运行模式取决于传递给 SparkContext 的 Master URL 的形式。Master URL 的形式有以下几种。

① local：使用一个 Worker 线程来本地化运行 Spark。

② local[*]：使用数量为逻辑 CPU 个数的线程来本地化运行 Spark。

③ local[K]：使用 *K* 个 Worker 线程来本地化运行 Spark（理想情况下，*K* 应该根据运行机器的 CPU 核心数设置）。

④ spark：//HOST:PORT 连接到指定的 Spark Standalone Master，默认端口是 7077。

⑤ yarn-client：以客户端模式连接 YARN 集群。集群的位置可以在 HADOOP_CONF_DIR 环境变量中找到。

⑥ yarn-cluster：以集群模式连接 YARN 集群。集群的位置可以在 HADOOP_CONF_DIR 环境变量中找到。

⑦ mesos://HOST:PORT 连接到指定的 Mesos 集群，默认端口是 5050。

在 Spark 中采用本地模式启动 PySpark 的命令主要包含以下参数。

① --master：表示当前的 PySpark 要连接到哪个 Master，如果是 local[*]，那么使用本地模式启动 PySpark，其中，中括号内的星号表示需要使用几个 CPU 核心，也就是启动几个线程来模拟 Spark 集群。

② --jars：用于把相关的 JAR 包添加到 CLASSPATH 中；如果有多个 JAR 包，可以使用逗号分隔符连接它们。

例如，要采用本地模式，在 4 个 CPU 核心上运行 PySpark，命令如下。

```
$cd  spark
$./bin/pyspark --master local[4]
```

或者，可以在 CLASSPATH 中添加 code.jar，命令如下。

```
$cd  spark
$./bin/pyspark --master local[4] --jars code.jar
```

可以执行 PySpark --help 命令，获取完整的选项列表，具体如下。

```
$cd  spark
$./bin/pyspark -help
```

执行以下命令启动 PySpark（默认是 local 模式）。

```
$cd  spark
$./bin/pyspark
```

启动 PySpark 成功后在输出信息的末尾可以看到>>>命令提示符，如图 6-16 所示。

图 6-16　启动 PySpark

输入 scala 代码进行调试，运行 scala 代码如图 6-17 所示。

可以使用命令"exit()"退出 PySpark，退出 PySpark 如图 6-18 所示。

图 6-17　运行 scala 代码

图 6-18　退出 PySpark

6.3.4　编程实现——Spark 独立应用程序

编写文件 WordCount.py，具体内容如下。

```
from pyspark import SparkConf, SparkContext
conf = SparkConf().setMaster("local").setAppName("My App")
sc = SparkContext(conf = conf)
logFile = "file:///usr/local/spark-3.3.0-bin-hadoop3/README.md"
logData = sc.textFile(logFile, 2).cache()
numAs = logData.filter(lambda line: 'a' in line).count()
numBs = logData.filter(lambda line: 'b' in line).count()
print('Lines with a: %s, Lines with b: %s' % (numAs, numBs))
```

① 对于这段 Python 代码，可以直接使用以下命令进行执行。

```
$cd  /usr/local/spark
$cd  /mycode/python
$python3 WordCount.py
```

执行上述命令以后得到以下结果。

```
Lines with a: 62, Lines with b: 30
```

② 通过 spark-submit 命令提交应用程序，该命令的格式如下。

```
spark-submit
  --master <master-url>
  --deploy-mode <deploy-mode>   # 部署模式
  ... # 其他参数
  <application-file>  # Python 代码文件
  [application-arguments]  # 传递给主类的主方法的参数
```

注：执行 spark-submit --help 命令可获取完整的选项列表，具体如下。

```
$./bin/spark-submit -help
```

通过 spark-submit 命令将应用程序提交到 Spark 中进行运行，命令如下。

```
$cd spark
$./bin/spark-submit  /home/hadoop/WordCount.py
```

在命令中间使用"\"符号，再按回车键，便可将一行完整命令人为地断开成多行，效果如下。

```
$/usr/local/spark/bin/spark-submit \
> /usr/local/mycode/WordCount.py
```

上面命令的执行结果如下。

```
Lines with a: 61, Lines with b: 30
```

为了避免其他多余信息对运行结果产生干扰，我们可以修改 log4j 的日志信息显示级别，具体如下。

```
cd /usr/local/spark/conf
cp log4j.properties.template log4j.properties
vim log4j.properties
```

待修改的命令如下。

```
log4j.arootCategory = INFO, console
```

修改后的命令如下。

```
log4j.rootCategory = ERROR, console
```

6.4 本章小结

本章主要介绍 Spark 的安装和配置方法，以及在 PySpark 环境中运行代码的方法，帮助读者更加深刻地理解 Spark 的使用方法。

第7章 Spark RDD：弹性分布式数据集

RDD不仅是Spark的核心概念，更是支撑Spark高效处理能力的关键组件。RDD代表了一个分布式、容错的数据集合，它的设计解决了大规模数据处理中最核心的问题，例如并行计算、容错性和数据分区。正是因为RDD，Spark得以实现出色的性能，具备灵活性，在大数据处理领域迅速崭露头角。此外，RDD的抽象设计也使开发人员能够更为轻松地编写并行和分布式代码，极大地推动了Spark的广泛采用。

【学习目标】

1．了解RDD及其使用场景。

2．掌握RDD常用算子的使用方法。

7.1 RDD概述

【任务描述】了解Spark RDD的概念、特性，理解RDD和传统数据集的区别。

1．RDD的概念和作用

RDD是一个弹性分布式数据集，表示一个分布在多台计算机上的、可以被并行处理的数据集合。它可以从稳定存储（例如HDFS、HBase等）中进行创建，或通过转换（例如映射、过滤等操作）从现有的RDD中进行创建，从而支持复杂的计算操作。

2．RDD在Spark中的关键作用

① 容错性：RDD在Spark中的一个重要作用是提供容错功能。当RDD一个分区的数据丢失时，Spark可以利用RDD的转换操作和依赖信息重新计算丢失的分区，从而实现数据的容错。这样，Spark在处理大规模数据时更加可靠。

② 并行计算：RDD可以分为多个分区，这些分区可以在集群的多台计算机上并行处理。这样，Spark可以在多台计算机上同时执行操作，加快数据的处理。

③ 内存计算：RDD允许将数据存储在内存中，从而提高数据访问速度。这样，Spark在某些情况下可实现比传统基于磁盘的计算框架更快的计算。

3．RDD的特性

① 不可变性：RDD是不可变的，这意味着RDD一旦创建，就不能进行修改。如果需要对RDD进行转换或其他操作，则其实质是创建一个新的RDD，而不是修改原始RDD。

不可变性有助于保持数据的一致性和可靠性。

② 依赖关系：每个 RDD 记录了自己与父 RDD 的依赖关系。这些依赖关系构成了一个有向无环图，允许 Spark 在数据丢失时重新计算所需的分区，从而实现容错功能。

③ 数据分区：RDD 可以分为多个数据分区，每个分区可以在集群的不同计算节点上并行处理。数据分区使数据处理可以在多个节点上同时进行，提高了计算效率。

④ 可持久化：RDD 可以将数据持久化存储到磁盘或内存中，以便在需要时进行使用。可持久化对于迭代计算和交互式数据分析来说非常有用。

4．RDD 与传统分布式数据集的区别

与传统的分布式数据集相比，RDD 具有更高的容错性和更灵活的操作能力。传统的分布式数据集通常需要通过复杂的序列化和反序列化操作，以在计算节点之间传递数据，而 RDD 通过将数据存储在内存中和利用转换操作，更高效地进行分布式计算。此外，RDD 的不可变性和依赖关系追踪使容错性的实现更加优雅。

7.2 RDD 编程

【任务描述】掌握 RDD 的操作方法、键值对 RDD 的操作方法，以及 RDD 对数据的读/写操作。

7.2.1 RDD 编程基础

1．创建 RDD

（1）从文件系统中加载数据创建 RDD

Spark 采用 textFile()方法从文件系统中加载数据来创建 RDD，该方法把文件的 URI 作为参数。这个 URI 可以是以下内容。

① 本地文件系统的地址。

② 分布式文件系统 HDFS 的地址。

③ Amazon S3 的地址。

使用 Python 创建 RDD 的示例如下。

从本地文件系统中加载数据来创建 RDD 的命令如下，示意如图 7-1 所示。

```
>>> from __future__ import print_function # 引入 print 函数
>>> lines = sc.textFile("file:///usr/local/spark/mycode/rdd/word.txt")
>>> lines.foreach(print)
Hadoop is good
Spark is fast
Spark is better
```

从分布式文件系统 HDFS 中加载数据，命令如下。

```
>>> lines = sc.textFile("hdfs:// localhost:9000/user/hadoop/word.txt")
>>> lines = sc.textFile("/user/hadoop/word.txt")
>>> lines = sc.textFile("word.txt")
```

上述 3 条语句的作用是完全等价的，读者可以使用其中任意一条语句。

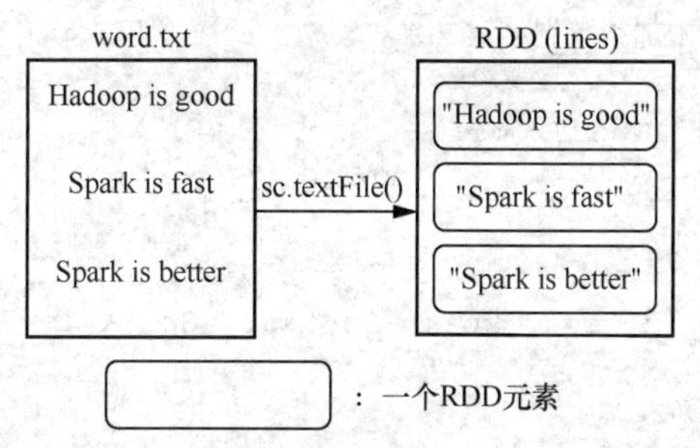

图 7-1　从本地文件系统中加载数据并创建 RDD 示意

（2）通过并行集合（列表）创建 RDD

我们可以调用 SparkContext 的 parallelize()方法，在 Driver 中一个已经存在的集合（列表）上创建 RDD。从数组创建 RDD 的命令如下，示意如图 7-2 所示。

```
>>> array = [1,2,3,4,5]
>>> rdd = sc.parallelize(array)
>>> rdd.foreach(print)
1
2
3
4
5
```

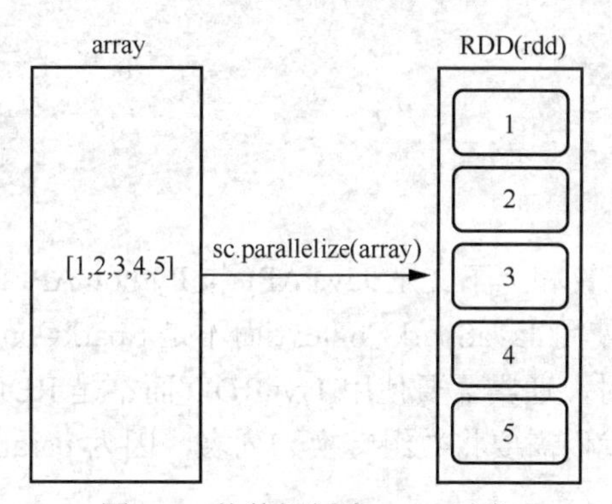

图 7-2　从数组创建 RDD 示意

使用 Java 创建 RDD 的示例代码如下。

```
import org.apache.spark.SparkConf;
import org.apache.spark.api.java.JavaRDD;
import org.apache.spark.api.java.JavaSparkContext;

public class SparkExample {
   public static void main(String[] args) {
       // 设置 Spark 配置
       SparkConf conf =
           new SparkConf().setAppName("SparkJavaExample").setMaster("local");
```

```
        JavaSparkContext sc = new JavaSparkContext(conf);

        // 从本地文件系统中加载数据创建 RDD
        // 加载本地文件
        JavaRDD<String> linesFromFile = sc.textFile(
                        "file:///usr/local/spark/mycode/rdd/word.txt");
        // 输出数据
        linesFromFile.foreach(line -> System.out.println(line));
        // 输出: Hadoop is good, Spark is fast, Spark is better

        // 从 HDFS 中加载数据
        JavaRDD<String> linesFromHDFS = sc.textFile("hdfs:// localhost:9000/user/
                        hadoop/word.txt");
        // 由于 HDFS 路径可以省略，以下两条语句与上面的语句效果相同
        // JavaRDD<String> linesFromHDFS = sc.textFile(
                        "/user/hadoop/word.txt");
        // JavaRDD<String> linesFromHDFS = sc.textFile("word.txt");

        // 通过并行集合（列表）创建 RDD
        Integer[] numbersArray = {1, 2, 3, 4, 5};
        JavaRDD<Integer> numbersRDD =
                    sc.parallelize(Arrays.asList(numbersArray));
        // 输出数据
        numbersRDD.foreach(num -> System.out.println(num));
        // 输出: 1, 2, 3, 4, 5

        // 关闭 Spark 上下文
        sc.close();
    }
}
```

注释解释如下。

① 初始化 Spark 的配置和上下文是 Java API 与 Python API 的主要区别。

② 在 Java 中，我们使用 JavaSparkContext 而不是 SparkContext。

③ 当使用 Java API 时，通常需要使用 JavaRDD 而不是 RDD。

④ 对于并行集合，我们需要将数组转换为列表，因为 parallelize()方法将一个集合作为参数。

2. RDD 操作

（1）转换操作

对于 RDD 来说，每次进行转换操作都会生成一个新的 RDD，为下一个“转换”操作提供数据。值得注意的是，这些通过转换得到的 RDD 采用的是惰性机制，这意味着转换过程中并不会立即进行数据计算，而是仅仅记录转换的步骤和路径。只有当遇到行动操作时，系统才会启动实际的数据计算。这时，数据计算会根据 RDD 的“血缘”关系从源头开始，按照转换的轨迹进行。RDD 转换操作如图 7-3 所示。常用的 RDD 转换操作如表 7-1 所示。

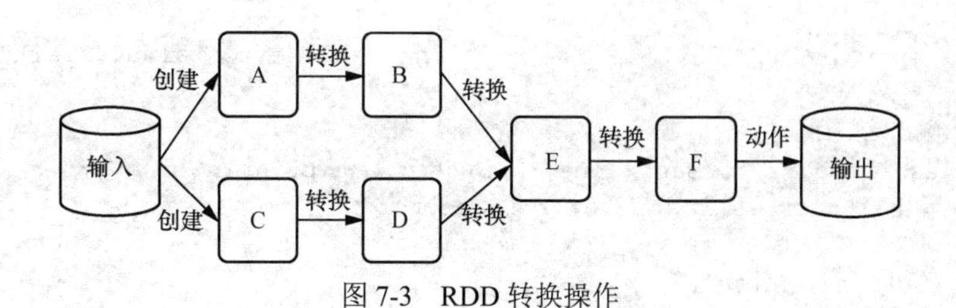

图 7-3　RDD 转换操作

表 7-1　常用的 RDD 转换操作

操作	含义
filter(func)	筛选出满足 func()函数的元素，并返回一个新的数据集
map(func)	将每个元素传递到 func()函数中，并将结果返回为一个新的数据集
flatMap(func)	与 map()函数相似，但每个输入元素都可以映射到 0 或多个输出结果上
groupByKey()	应用于键值对数据集时，返回一个新的(K, Iterable)形式的数据集，其中，K 表示键（key），Iterable 表示可迭代对象
reduceByKey(func)	应用于键值对数据集时，返回一个新的(K, V)形式的数据集，其中，K 表示键（key），V 表示值（Value）。每个值是将每个键传递到 func()函数中进行聚合后的结果

① filter(func)

筛选出满足 func()函数的元素，并返回一个新的数据集。

Python 示例代码如下。

```
>>> lines = sc.textFile("file:///usr/local/spark/mycode/rdd/word.txt")
>>> linesWithSpark = lines.filter(lambda line: "Spark" in line)
>>> linesWithSpark.foreach(print)
Spark is better
Spark is fast
```

Java 示例代码如下。

```
import org.apache.spark.SparkConf;
import org.apache.spark.api.java.JavaRDD;
import org.apache.spark.api.java.JavaSparkContext;

public class SparkJavaFilterExample {
    public static void main(String[] args) {
        // 设置 Spark 配置
        SparkConf conf = new SparkConf().setAppName(
                    "SparkJavaFilterExample").setMaster("local");
        JavaSparkContext sc = new JavaSparkContext(conf);

        // 从本地文件系统中加载数据创建 RDD
        JavaRDD<String> lines = sc.textFile(
                        "file:///usr/local/spark/mycode/rdd/word.txt");

        // 过滤包含"Spark"的行
        JavaRDD<String> linesWithSpark = lines.filter(
```

```
                                        line -> line.contains("Spark"));

        // 输出这些行
        linesWithSpark.foreach(line -> System.out.println(line));
        // 输出: Spark is fast, Spark is better

        // 关闭 Spark 上下文
        sc.close();
    }
}
```

在这个 Java 示例代码中，我们使用了 Java 的 lambda 表达式代替 Python 的 lambda 函数；过滤逻辑是基于字符串的 contains()方法实现的。filter()操作实例执行过程如图 7-4 所示。

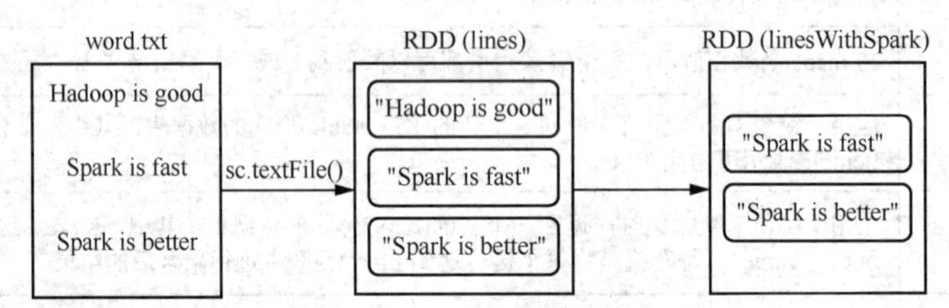

图 7-4　filter()操作实例执行过程

② map(func)

map(func)是一个转换操作，它将数据集中的每个元素应用于 func()函数，然后生成一个包含处理结果的新数据集。map()操作对通过数组创建 RDD 的执行过程如图 7-5 所示。

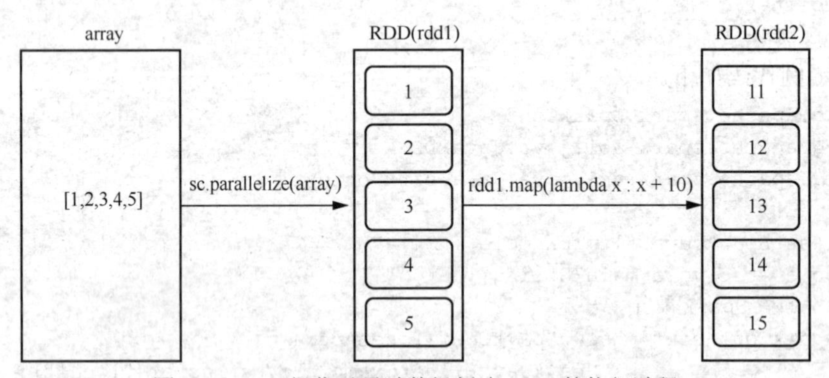

图 7-5　map()操作对通过数组创建 RDD 的执行过程

Python 示例代码如下。

```
>>> array = [1,2,3,4,5]
>>> rdd1 = sc.parallelize(array)
>>> rdd2 = rdd1.map(lambda x:x+10)
>>> rdd2.foreach(print)
11
```

```
13
12
14
15
```

Java 示例代码如下。

```
import org.apache.spark.SparkConf;
import org.apache.spark.api.java.JavaRDD;
import org.apache.spark.api.java.JavaSparkContext;
import org.apache.spark.api.java.function.Function;
import org.apache.spark.api.java.function.VoidFunction;

import java.util.Arrays;
import java.util.List;

public class MapExample {
    public static void main(String[] args) {
        // 初始化 Spark 配置和上下文
        SparkConf conf =
            new SparkConf().setAppName("Java Map Example").setMaster("local");
        JavaSparkContext sc = new JavaSparkContext(conf);

        // 创建一个数据集合
        List<Integer> data = Arrays.asList(1, 2, 3, 4, 5);

        // 创建一个 Java RDD
        JavaRDD<Integer> rdd1 = sc.parallelize(data);

        // 使用 map()函数进行转换操作
        JavaRDD<Integer> rdd2 = rdd1.map(new Function<Integer, Integer>() {
            @Override
            public Integer call(Integer x) throws Exception {
                return x + 10;
            }
        });

        // 打印转换后的数据
        rdd2.foreach(new VoidFunction<Integer>() {
            @Override
            public void call(Integer x) throws Exception {
                System.out.println(x);
            }
        });

        // 关闭 Spark 上下文
        sc.close();
    }
}
```

map()操作对从本地文件系统中加载数据集创建 RDD 的执行过程如图 7-6 所示。

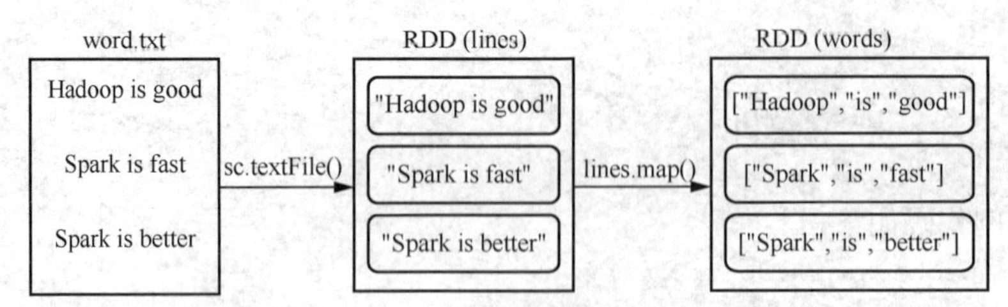

图 7-6　map()操作对从本地文件系统中加载数据集创建 RDD 的执行过程

Python 示例代码如下。

```
>>> lines = sc.textFile("file:///usr/local/spark/mycode/rdd/word.txt")
>>> words = lines.map(lambda line:line.split(" "))
>>> words.foreach(print)
['Hadoop', 'is', 'good']
['Spark', 'is', 'fast']
['Spark', 'is', 'better']
```

Java 示例代码如下。

```
import org.apache.spark.SparkConf;
import org.apache.spark.api.java.JavaRDD;
import org.apache.spark.api.java.JavaSparkContext;
import org.apache.spark.api.java.function.Function;
import org.apache.spark.api.java.function.VoidFunction;

import java.util.Arrays;
import java.util.List;

public class SplitWordsExample {
    public static void main(String[] args) {
        // 初始化 Spark 配置和上下文
        SparkConf conf = new SparkConf().setAppName("Java Split Words
Example").setMaster("local");
        JavaSparkContext sc = new JavaSparkContext(conf);

        // 从文本文件中读取行
        JavaRDD<String> lines = sc.textFile(
                                "file:///usr/local/spark/mycode/rdd/word.txt");

        // 使用 map()函数将每行文本分割为单词
        JavaRDD<List<String>> words = lines.map(new Function<String,
                                        List<String>>() {
            @Override
            public List<String> call(String line) throws Exception {
                return Arrays.asList(line.split(" "));
            }
        });

        // 打印分割后的单词列表
        words.foreach(new VoidFunction<List<String>>() {
            @Override
```

```
            public void call(List<String> wordList) throws Exception {
                System.out.println(wordList);
            }
        });

        // 关闭 Spark 上下文
        sc.close();
    }
}
```

③ flatMap(func)

flatMap(func)操作首先将 map()函数应用于 RDD 的所有元素，然后将结果平坦化，最后返回新的 RDD。flatMap()操作实例执行过程如图 7-7 所示，示例代码如下。

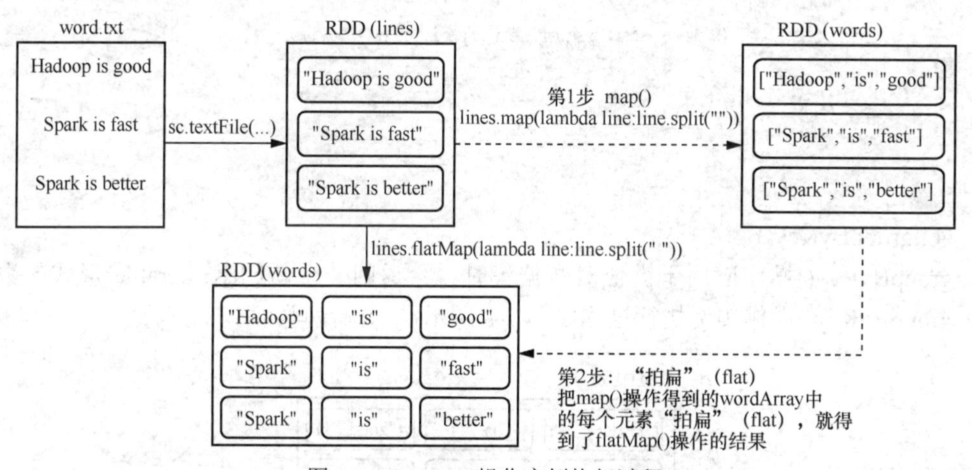

图 7-7 flatMap()操作实例执行过程

Python 示例代码如下。

```
>>> lines = sc.textFile("file:///usr/local/spark/mycode/rdd/word.txt")
>>> words = lines.flatMap(lambda line:line.split(" "))
```

Java 示例代码如下。

```
import org.apache.spark.SparkConf;
import org.apache.spark.api.java.JavaRDD;
import org.apache.spark.api.java.JavaSparkContext;
import org.apache.spark.api.java.function.FlatMapFunction;

import java.util.Arrays;
import java.util.Iterator;

public class FlatMapWordsExample {
    public static void main(String[] args) {
        // 初始化 Spark 配置和上下文
        SparkConf conf = new SparkConf().setAppName("Java FlatMap Words
                        Example").setMaster("local");
        JavaSparkContext sc = new JavaSparkContext(conf);
```

```
        // 从文本文件中读取行
        JavaRDD<String> lines = sc.textFile(
                                "file:///usr/local/spark/mycode/rdd/word.txt");

        // 使用 flatMap()函数将每行文本分割为单词，并将所有单词放在一个扁平的列表中
        JavaRDD<String> words = lines.flatMap(new FlatMapFunction<String,
                                String>() {
            @Override
            public Iterator<String> call(String line) throws Exception {
                return Arrays.asList(line.split(" ")).iterator();
            }
        });

        // 打印每个单词
        words.foreach(word -> System.out.println(word));

        // 关闭 Spark 上下文
        sc.close();
    }
}
```

④ groupByKey()

groupByKey()操作应用于键值对数据集时，会返回一个新的(K, Iterable)形式的数据集。groupByKey()操作实例执行过程如图 7-8 所示，示例代码如下。

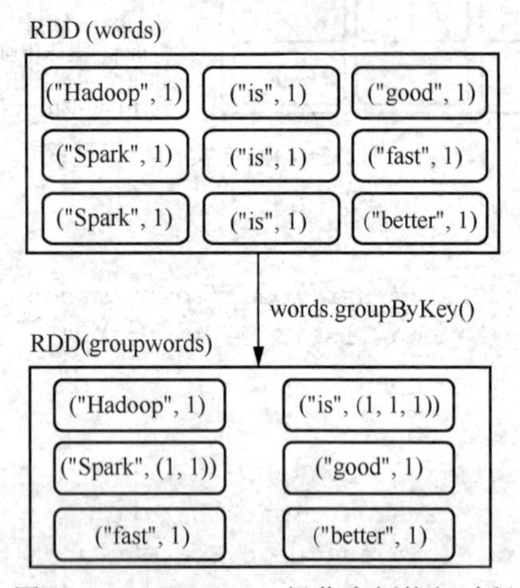

图 7-8　groupByKey()操作实例执行过程

Python 示例代码如下。

```
>>> words = sc.parallelize([("Hadoop",1),("is",1),("good",1), \
... ("Spark",1),("is",1),("fast",1),("Spark",1),("is",1),("better",1)])
>>> words1 = words.groupByKey()
>>> words1.foreach(print)
('Hadoop', <pyspark.resultiterable.ResultIterable object at 0x7fb210552c88>)
('better', <pyspark.resultiterable.ResultIterable object at 0x7fb210552e80>)
('fast', <pyspark.resultiterable.ResultIterable object at 0x7fb210552c88>)
```

```
('good', <pyspark.resultiterable.ResultIterable object at 0x7fb210552c88>)
('Spark', <pyspark.resultiterable.ResultIterable object at 0x7fb210552f98>)
('is', <pyspark.resultiterable.ResultIterable object at 0x7fb210552e10>)
```

Java 示例代码如下。

```
import org.apache.spark.SparkConf;
import org.apache.spark.api.java.JavaPairRDD;
import org.apache.spark.api.java.JavaSparkContext;
import scala.Tuple2;

import java.util.Arrays;
import java.util.List;

public class GroupByKeyExample {
    public static void main(String[] args) {
        // 初始化 Spark 配置和上下文
        SparkConf conf = new SparkConf().setAppName(
                "Java GroupByKey Example").setMaster("local");
        JavaSparkContext sc = new JavaSparkContext(conf);

        // 创建一个键值对数据集
        List<Tuple2<String, Integer>> wordPairs = Arrays.asList(
                new Tuple2<>("Hadoop", 1),
                new Tuple2<>("is", 1),
                new Tuple2<>("good", 1),
                new Tuple2<>("Spark", 1),
                new Tuple2<>("is", 1),
                new Tuple2<>("fast", 1),
                new Tuple2<>("Spark", 1),
                new Tuple2<>("is", 1),
                new Tuple2<>("better", 1)
        );
        JavaPairRDD<String, Integer> words = sc.parallelizePairs(wordPairs);

        // 使用 groupByKey()函数
        JavaPairRDD<String, Iterable<Integer>> words1 = words.groupByKey();

        // 打印结果
        words1.foreach(word -> {
            System.out.println("(" + word._1 + ", " + word._2 + ")");
        });

        // 关闭 Spark 上下文
        sc.close();
    }
}
```

注意，在 Java 示例代码的输出中，每个键的值将被表示为一个迭代器，这与 Python 中的'ResultIterable'类似。但在 Python 示例代码中，输出会被直接显示为一个列表。

⑤ reduceByKey(func)

reduceByKey(func)操作应用于键值对数据集时，会返回一个新的(K, V)形式的数据集。reduceByKey()操作实例执行过程如图 7-9 所示，示例代码如下。

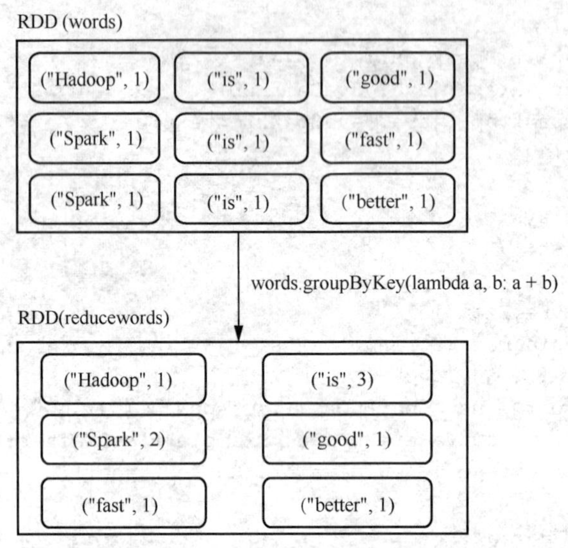

图 7-9　reduceByKey()操作实例执行过程

Python 示例代码如下。

```
>>> words = sc.parallelize([("Hadoop",1),("is",1),("good",1),("Spark",1), \
... ("is",1),("fast",1),("Spark",1),("is",1),("better",1)])
>>> words1 = words.reduceByKey(lambda a,b:a+b)
>>> words1.foreach(print)
('good', 1)
('Hadoop', 1)
('better', 1)
('Spark', 2)
('fast', 1)
('is', 3)
```

Java 示例代码如下。

```
import org.apache.spark.SparkConf;
import org.apache.spark.api.java.JavaPairRDD;
import org.apache.spark.api.java.JavaSparkContext;
import org.apache.spark.api.java.function.Function2;
import scala.Tuple2;

import java.util.Arrays;
import java.util.List;

public class ReduceByKeyExample {
    public static void main(String[] args) {
        // 初始化 Spark 配置和上下文
        SparkConf conf = new SparkConf().setAppName(
                        "Java ReduceByKey Example").setMaster("local");
```

```
        JavaSparkContext sc = new JavaSparkContext(conf);

        // 创建一个键值对数据集
        List<Tuple2<String, Integer>> wordPairs = Arrays.asList(
                new Tuple2<>("Hadoop", 1),
                new Tuple2<>("is", 1),
                new Tuple2<>("good", 1),
                new Tuple2<>("Spark", 1),
                new Tuple2<>("is", 1),
                new Tuple2<>("fast", 1),
                new Tuple2<>("Spark", 1),
                new Tuple2<>("is", 1),
                new Tuple2<>("better", 1)
        );
        JavaPairRDD<String, Integer> words = sc.parallelizePairs(wordPairs);

        // 使用 reduceByKey()函数
        JavaPairRDD<String, Integer> words1 =
                words.reduceByKey(new Function2<Integer, Integer, Integer>() {
            @Override
            public Integer call(Integer a, Integer b) throws Exception {
                return a + b;
            }
        });

        // 打印结果
        words1.foreach(word -> {
            System.out.println("(" + word._1 + ", " + word._2 + ")");
        });

        // 关闭 Spark 上下文
        sc.close();
    }
}
```

（2）行动操作

行动操作会真正触发计算。Spark 程序执行到行动操作时，才会执行真正的计算，即从文件中加载数据，完成一次又一次转换操作，最终得到结果。常用的 RDD 行动操作如表 7-2 所示。

表 7-2　常用的 RDD 行动操作

操作	含义
count()	返回数据集中元素的个数
collect()	以数组的形式返回数据集的所有元素
first()	返回数据集的第一个元素

续表

操作	含义
take(n)	以数组的形式返回数据集的前 *n* 个元素
reduce(func)	通过 func()函数（输入两个参数并返回一个值）聚合数据集的元素
foreach(func)	将数据集的每个元素传递到 func()函数中运行

RDD 行动操作的 Python 示例代码如下。

```
>>> rdd = sc.parallelize([1,2,3,4,5])
>>> rdd.count()
5
>>> rdd.first()
1
>>> rdd.take(3)
[1, 2, 3]
>>> rdd.reduce(lambda a,b:a+b)
15
>>> rdd.collect()
[1, 2, 3, 4, 5]
>>> rdd.foreach(lambda elem:print(elem))
1
2
3
4
5
```

RDD 行动操作的 Java 示例代码如下。

```
import org.apache.spark.SparkConf;
import org.apache.spark.api.java.JavaRDD;
import org.apache.spark.api.java.JavaSparkContext;
import org.apache.spark.api.java.function.Function2;
import org.apache.spark.api.java.function.VoidFunction;

import java.util.Arrays;
import java.util.List;

public class RDDOperationsExample {
    public static void main(String[] args) {
        // 初始化 Spark 配置和上下文
        SparkConf conf = new SparkConf().setAppName(
                    "Java RDD Operations Example").setMaster("local");
        JavaSparkContext sc = new JavaSparkContext(conf);

        // 创建 RDD
        JavaRDD<Integer> rdd = sc.parallelize(Arrays.asList(1, 2, 3, 4, 5));

        // 使用 count()操作
        System.out.println("Count: " + rdd.count());
```

```
        // 使用 first()操作
        System.out.println("First: " + rdd.first());

        // 使用 take()操作
        List<Integer> firstThree = rdd.take(3);
        System.out.println("Take(3): " + firstThree);

        // 使用 reduce()操作
        int sum = rdd.reduce(new Function2<Integer, Integer, Integer>() {
            @Override
            public Integer call(Integer a, Integer b) throws Exception {
                return a + b;
            }
        });
        System.out.println("Sum: " + sum);

        // 使用 collect()操作
        List<Integer> collected = rdd.collect();
        System.out.println("Collect: " + collected);

        // 使用 foreach()操作
        rdd.foreach(new VoidFunction<Integer>() {
            @Override
            public void call(Integer elem) throws Exception {
                System.out.println(elem);
            }
        });

        // 关闭 Spark 上下文
        sc.close();
    }
}
```

（3）惰性机制

惰性机制是指整个转换过程中只记录转换的轨迹，并不会发生真正的计算，只有遇到行动操作时，才会触发“从头到尾”的真正的计算。

下面通过一个简单的示例来展示 Spark 的惰性机制。

Python 示例代码如下。

```
>>> lines = sc.textFile("file:///usr/local/spark/mycode/rdd/word.txt")
>>> lineLengths = lines.map(lambda s:len(s))
>>> totalLength = lineLengths.reduce(lambda a,b:a+b)
>>> print(totalLength)
42
```

Java 示例代码如下。

```
import org.apache.spark.SparkConf;
import org.apache.spark.api.java.JavaRDD;
import org.apache.spark.api.java.JavaSparkContext;
```

```
import org.apache.spark.api.java.function.Function;
import org.apache.spark.api.java.function.Function2;

public class LineLengthExample {
    public static void main(String[] args) {
        // 初始化 Spark 配置和上下文
        SparkConf conf = new SparkConf().setAppName(
                        "Java Line Length Example").setMaster("local");
        JavaSparkContext sc = new JavaSparkContext(conf);

        // 读取文本文件
        JavaRDD<String> lines = sc.textFile(
                                "file:///usr/local/spark/mycode/rdd/word.txt");

        // 使用 map() 操作计算每一行的长度
        JavaRDD<Integer> lineLengths = lines.map(new Function<String,
                                                    Integer>() {
            @Override
            public Integer call(String s) throws Exception {
                return s.length();
            }
        });

        // 使用 reduce() 操作计算所有行的总长度
        int totalLength = lineLengths.reduce(new Function2<Integer,
                                            Integer, Integer>() {
            @Override
            public Integer call(Integer a, Integer b) throws Exception {
                return a + b;
            }
        });

        // 打印结果
        System.out.println(totalLength);

        // 关闭 Spark 上下文
        sc.close();
    }
}
```

可以看出在 Spark 中，RDD 每次遇到行动操作，都会从头开始执行计算。每次调用行动操作，都会触发一次从头开始的计算。这对于迭代计算而言，代价是很大的，迭代计算经常需要多次重复使用同一组数据。

以下是多次计算同一个 RDD 的例子。

Python 示例代码如下。

```
>>> list = ["Hadoop","Spark","Hive"]
>>> rdd = sc.parallelize(list)
>>> print(rdd.count()) // 行动操作，触发一次真正从头到尾的计算
3
```

```
>>> print(','.join(rdd.collect())) // 行动操作，触发一次真正从头到尾的计算
Hadoop,Spark,Hive
```

Java 示例代码如下。

```
import org.apache.spark.SparkConf;
import org.apache.spark.api.java.JavaRDD;
import org.apache.spark.api.java.JavaSparkContext;

import java.util.Arrays;
import java.util.List;
import java.util.stream.Collectors;

public class RDDOperationsJava {
    public static void main(String[] args) {
        // 初始化 Spark 配置和上下文
        SparkConf conf = new SparkConf().setAppName("Java RDD
                    Operations").setMaster("local");
        JavaSparkContext sc = new JavaSparkContext(conf);

        // 创建一个列表并转换为 RDD
        List<String> dataList = Arrays.asList("Hadoop", "Spark", "Hive");
        JavaRDD<String> rdd = sc.parallelize(dataList);

        // 使用 count()操作，并打印结果
        System.out.println(rdd.count());

        // 使用 collect()操作，并使用 Java 8 的 Stream API 将结果转换为一个字符串
        String result = String.join(",", rdd.collect());
        System.out.println(result);

        // 关闭 Spark 上下文
        sc.close();
    }
}
```

3. 数据持久化

Spark 有一个很重要的能力，就是将数据持久化，在多个操作间可以访问这些持久化的数据。当持久化一个 RDD 时，每个节点的其他分区都可以使用 RDD 在内存中进行计算，在该数据上的其他行动操作将直接使用内存中的数据。这样会让以后的行动操作计算速度加快（通常运行速度会快 10 倍）。数据持久化是迭代算法和快速交互式使用的重要工具。

持久化（缓存）机制避免这种重复计算的开销，使用 persist()方法对一个 RDD 标记为持久化。之所以说“标记为持久化”，是因为出现 persist()语句的地方并不会马上计算生成 RDD 并把它持久化，而是要等到第一个行动操作触发真正计算以后，才会对计算结果进行持久化。持久化后的 RDD 将会被保留在计算节点的内存中，被后面的行动操作重复使用。

persist()的圆括号中包含的是持久化级别参数。

① persist(MEMORY_ONLY)表示将 RDD 作为反序列化的对象存储于 JVM 中，如果内存不足，就要按照最近最少使用（Least Recently Used，LRU）原则替换缓存中的内容。

② persist(MEMORY_AND_DISK)表示将 RDD 作为反序列化的对象存储在 JVM 中，如果内存不足，超出的分区将会被存储在硬盘上。

③ 一般而言，使用 cache()方法时，会调用 persist(MEMORY_ONLY)，可以使用 unpersist()方法手动地把持久化的 RDD 从缓存中移除。

针对上面的实例，增加持久化语句以后的执行过程如下。

Python 示例代码如下。

```
>>> list = ["Hadoop","Spark","Hive"]
>>> rdd = sc.parallelize(list)
>>> rdd.cache()
# 会调用 persist(MEMORY_ONLY)，但是，语句执行到这里，并不会缓存 RDD，因为这时 RDD 还没有被
# 计算生成
>>> print(rdd.count())
# 第一次行动操作，触发一次真正从头到尾的计算，这时上面的 rdd.cache()才会被执行，把这个 RDD
# 存储到缓存中
3
>>> print(','.join(rdd.collect()))
# 第二次行动操作，不需要触发从头到尾的计算，只需要重复使用上面缓存中的 RDD
Hadoop,Spark,Hive
```

Java 示例代码如下。

```
import org.apache.spark.SparkConf;
import org.apache.spark.api.java.JavaRDD;
import org.apache.spark.api.java.JavaSparkContext;
import org.apache.spark.storage.StorageLevel;

import java.util.Arrays;
import java.util.List;

public class RDDCachingExample {
    public static void main(String[] args) {
        // 初始化 Spark 配置和上下文
        SparkConf conf = new SparkConf().setAppName(
                        "Java RDD Caching Example").setMaster("local");
        JavaSparkContext sc = new JavaSparkContext(conf);

        // 创建一个列表并转换为 RDD
        List<String> dataList = Arrays.asList("Hadoop", "Spark", "Hive");
        JavaRDD<String> rdd = sc.parallelize(dataList);

        // 标记 RDD 为缓存
        rdd.persist(StorageLevel.MEMORY_ONLY());  // 这等同于 rdd.cache()

        // 第一次行动操作
        System.out.println(rdd.count());

        // 第二次行动操作
        System.out.println(String.join(",", rdd.collect()));

        // 关闭 Spark 上下文
```

```
        sc.close();
    }
}
```

4．分区

RDD 是弹性分布式数据集，通常很大，会被分成很多个分区，分别保存在不同的节点上。

（1）分区的作用

① 增加并行度，RDD 分区被保存到不同节点上，如图 7-10 所示。

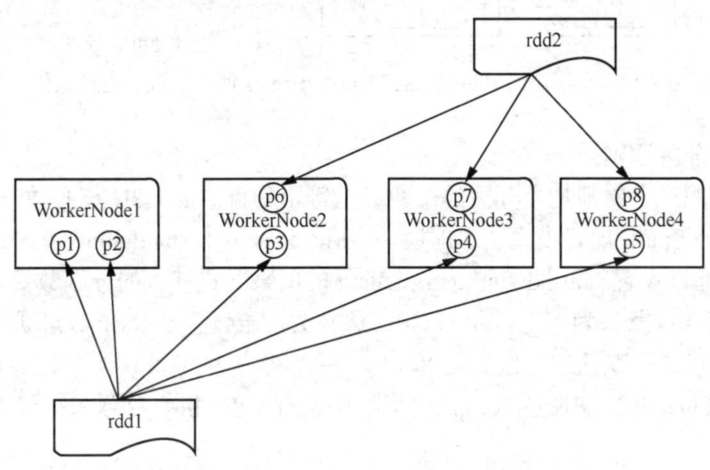

图 7-10　RDD 分区被保存到不同节点上

② 减少通信开销

现有 UserData（UserId，UserInfo）、Events（UserID，LinkInfo），对 UserData 和 Events 表进行连接操作，获得（UserID，UserInfo，LinkInfo），未分区时对 UserData 和 Events 两个表进行连接操作如图 7-11 所示，采用分区以后对 UserData 和 Events 两个表进行连接操作如图 7-12 所示。

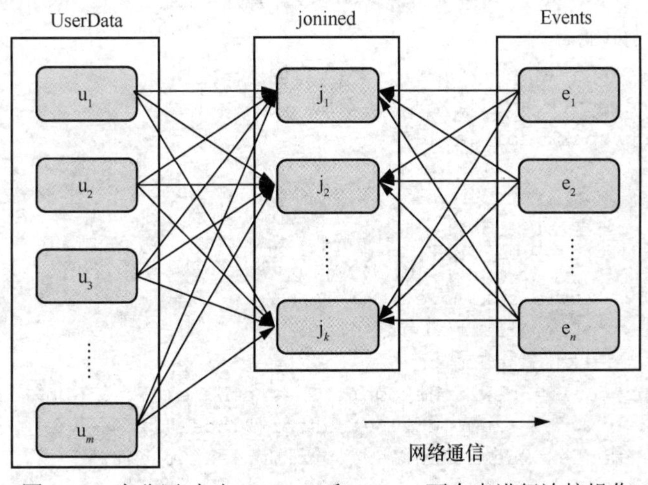

图 7-11　未分区时对 UserData 和 Events 两个表进行连接操作

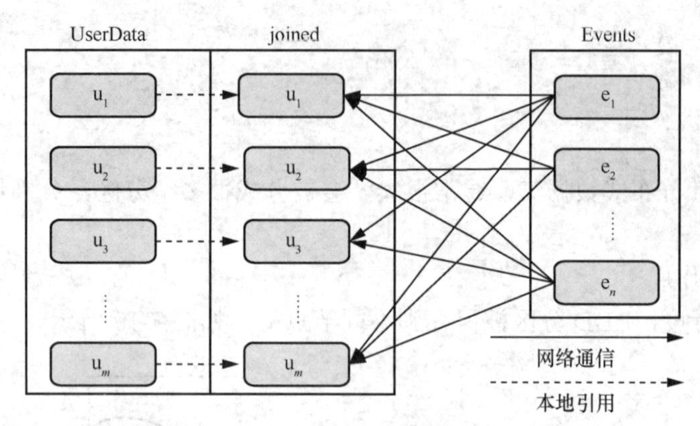

图 7-12　采用分区以后对 UserData 和 Events 两个表进行连接操作

（2）RDD 分区原则

RDD 分区的一个原则是使分区的个数尽量等于集群中的 CPU 核心数。

对于不同的 Spark 部署模式（本地模式、Mesos 模式、Standalone 模式、YARN 模式）而言，都可以通过设置 spark.default.parallelism 的值来配置默认的分区数目，具体如下。

① 本地模式：默认为本地机器的 CPU 核心数，若设置了 local[N]，则默认为 N。

② Mesos 模式：默认的分区数为 8。

③ Standalone 或 YARN 模式：在“集群中所有 CPU 核心数总和”和“2”之中取较大值作为默认值。

（3）设置分区的个数

① 创建 RDD 时手动指定分区个数。

在调用 textFile()和 parallelize()方法时，可以手动指定分区个数，语法格式如下。

```
sc.textFile(path, partitionNum)
```

其中，path 参数用于指定要加载的文件的地址，partitionNum 参数用于指定分区个数。

Python 示例代码如下。

```
>>> list = [1,2,3,4,5]
>>> rdd = sc.parallelize(list,2)  # 设置两个分区
```

Java 示例代码如下。

```
import org.apache.spark.SparkConf;
import org.apache.spark.api.java.JavaRDD;
import org.apache.spark.api.java.JavaSparkContext;

import java.util.Arrays;
import java.util.List;

public class RDDPartitionsExample {
    public static void main(String[] args) {
        // 初始化 Spark 配置和上下文
        SparkConf conf = new SparkConf().setAppName("Java RDD Partitions
                     Example").setMaster("local");
        JavaSparkContext sc = new JavaSparkContext(conf);
```

```
        // 创建一个列表并转换为 RDD，设置 2 个分区
        List<Integer> dataList = Arrays.asList(1, 2, 3, 4, 5);
        JavaRDD<Integer> rdd = sc.parallelize(dataList, 2);

        // 现在，你可以对这个 RDD 执行其他操作
        // 例如：打印每个元素
        rdd.foreach(element -> System.out.println(element));

        // 关闭 Spark 上下文
        sc.close();
    }
}
```

② 使用 repartition()方法重新设置分区个数。

通过转换操作得到新 RDD，直接调用 repartition()方法即可。

Python 示例代码如下。

```
>>> data = sc.parallelize([1,2,3,4,5],2)
>>> len(data.glom().collect())   # 显示 data 这个 RDD 的分区数量
2
>>> rdd = data.repartition(1)    # 对 data 这个 RDD 进行重新分区
>>> len(rdd.glom().collect())    # 显示 rdd 这个 RDD 的分区数量
1
```

Java 示例代码如下。

```
import org.apache.spark.SparkConf;
import org.apache.spark.api.java.JavaRDD;
import org.apache.spark.api.java.JavaSparkContext;

import java.util.Arrays;
import java.util.List;

public class RDDRepartitionExample {
    public static void main(String[] args) {
        // 初始化 Spark 配置和上下文
        SparkConf conf = new SparkConf().setAppName("Java RDD Repartition
                         Example").setMaster("local");
        JavaSparkContext sc = new JavaSparkContext(conf);

        // 创建一个 RDD 并设置 2 个分区
        List<Integer> dataList = Arrays.asList(1, 2, 3, 4, 5);
        JavaRDD<Integer> data = sc.parallelize(dataList, 2);

        // 显示 data 这个 RDD 的分区数量
        System.out.println(data.glom().collect().size());

        // 对 data 这个 RDD 进行重新分区
        JavaRDD<Integer> rdd = data.repartition(1);

        // 显示 rdd 这个 RDD 的分区数量
        System.out.println(rdd.glom().collect().size());
```

```
        // 关闭 Spark 上下文
        sc.close();
    }
}
```

（4）自定义分区方法

Spark 提供了自带的 HashPartitioner（哈希分区）与 RangePartitioner（区域分区），能够满足大多数应用场景的需求。与此同时，Spark 也支持自定义分区方式，即通过提供一个自定义的分区函数来控制 RDD 的分区方式，从而利用领域知识进一步减少通信开销。

实例：根据键的最后一位数字，将数据写入不同的文件。

例如：

```
10 写入到 part-00000
11 写入到 part-00001
...
19 写入到 part-00009
```

Python 示例代码如下。

```
from pyspark import SparkConf, SparkContext
def MyPartitioner(key):
    print("MyPartitioner is running")
    print('The key is %d' % key)
    return key%10

def main():
    print("The main function is running")
    conf  =  SparkConf().setMaster("local").setAppName("MyApp")
    sc  =  SparkContext(conf = conf)
    data = sc.parallelize(range(10),5)
    data.map(lambda x:(x,1)) \
           .partitionBy(10,MyPartitioner) \
           .map(lambda x:x[0]) \
           .saveAsTextFile("file:///usr/local/spark/mycode/rdd/partitioner")

if __name__ == '__main__':
       main()
```

使用以下命令运行 TestPartitioner.py 文件。

```
$cd /usr/local/spark/mycode/rdd
$/usr/local/spark/bin/spark-submit TestPartitioner.py
```

Java 示例代码如下。

```
import org.apache.spark.Partitioner;
import org.apache.spark.SparkConf;
import org.apache.spark.api.java.JavaPairRDD;
import org.apache.spark.api.java.JavaRDD;
import org.apache.spark.api.java.JavaSparkContext;
import scala.Tuple2;

import java.util.List;
```

```
import java.util.stream.Collectors;
import java.util.stream.IntStream;

public class TestPartitionerJava {

    public static class MyJavaPartitioner extends Partitioner {
        private int numPartitions;

        public MyJavaPartitioner(int numPartitions) {
            this.numPartitions = numPartitions;
        }

        @Override
        public int numPartitions() {
            return this.numPartitions;
        }

        @Override
        public int getPartition(Object key) {
            return ((int) key) % 10;
        }
    }

    public static void main(String[] args) {
        SparkConf conf = new SparkConf().setAppName("MyJavaApp").setMaster("local");
        JavaSparkContext sc = new JavaSparkContext(conf);

        List<Integer> data = IntStream.range(0, 10).boxed().collect(
                             Collectors.toList());
        JavaRDD<Integer> rdd = sc.parallelize(data, 5);

        JavaPairRDD<Integer, Integer> pairRdd = rdd.mapToPair(x -> new
                                                Tuple2<>(x, 1));
        JavaPairRDD<Integer, Integer> partitionedRdd = pairRdd.partitionBy(
                                                       new MyJavaPartitioner(10));

        JavaRDD<Integer> result = partitionedRdd.map(Tuple2::_1);
        result.saveAsTextFile(
            "file:///usr/local/spark/mycode/rdd/partitionerJava");

        sc.close();
    }
}
```

程序运行结果如下。

```
The main function is running
MyPartitioner is running
The key is 0
MyPartitioner is running
The key is 1
```

```
...
MyPartitioner is running
The key is 9
```

7.2.2 键值对 RDD

1．键值对 RDD 的创建

（1）第一种创建方式：从文件中加载

创建键值对 RDD 的方式有多种，其中主要的方式是使用 map()函数。实现读取文件内容，进行词频统计功能的示例代码如下。

Python 示例代码如下。

```
>>> lines = sc.textFile("file:// /usr/local/spark/mycode/pairrdd/word.txt")
>>> pairRDD = lines.flatMap(lambda line:line.split(" ")).map(lambda word:(word,1))
>>> pairRDD.foreach(print)
('I', 1)
('love', 1)
('Hadoop', 1)
...
```

Java 示例代码如下。

```
import org.apache.spark.SparkConf;
import org.apache.spark.api.java.JavaPairRDD;
import org.apache.spark.api.java.JavaRDD;
import org.apache.spark.api.java.JavaSparkContext;
import scala.Tuple2;

public class PairRDDFromTextFileJava {

    public static void main(String[] args) {
        SparkConf conf = new SparkConf().setAppName(
                       "PairRDDFromTextFileJavaApp").setMaster("local");
        JavaSparkContext sc = new JavaSparkContext(conf);

        JavaRDD<String> lines = sc.textFile(
                              "file:// /usr/local/spark/mycode/pairrdd/word.txt");

        JavaPairRDD<String, Integer> pairRDD = lines
            .flatMap(line -> java.util.Arrays.asList(
                 line.split(" ")).iterator())
            .mapToPair(word -> new Tuple2<>(word, 1));

        pairRDD.foreach(tuple -> System.out.println("(" + tuple._1() + "," +
                        tuple._2() + ")"));

        sc.close();
    }
}
```

（2）第二种创建方式：通过并行集合（列表）创建 RDD

实现并行化列表创建 RDD，将列表中每个元素标记为<元素, 1>的示例代码如下。

Python 示例代码如下。

```
>>> list = ["Hadoop","Spark","Hive","Spark"]
>>> rdd = sc.parallelize(list)
>>> pairRDD = rdd.map(lambda word:(word,1))
>>> pairRDD.foreach(print)
(Hadoop,1)
(Spark,1)
(Hive,1)
(Spark,1)
```

Java 示例代码如下。

```
import org.apache.spark.SparkConf;
import org.apache.spark.api.java.JavaPairRDD;
import org.apache.spark.api.java.JavaRDD;
import org.apache.spark.api.java.JavaSparkContext;
import scala.Tuple2;
import java.util.Arrays;
import java.util.List;


public class CreatePairRDDJava {
    public static void main(String[] args) {
        // 初始化 Spark 配置和上下文
        SparkConf conf = new SparkConf().setAppName(
                    "CreatePairRDDJavaApp").setMaster("local");
        JavaSparkContext sc = new JavaSparkContext(conf);

        // 创建一个 Java RDD
        List<String> list = Arrays.asList("Hadoop", "Spark", "Hive", "Spark");
        JavaRDD<String> rdd = sc.parallelize(list);

        // 从 RDD 创建一个键值对 RDD
        JavaPairRDD<String, Integer> pairRDD = rdd.mapToPair(
                                                word -> new Tuple2<>(word, 1));

        // 打印键值对 RDD 的内容
        pairRDD.foreach(tuple -> System.out.println("(" + tuple._1() + "," +
        tuple._2() + ")"));

        // 关闭 Spark 上下文
        sc.close();
    }
}
```

2. 常用的键值对 RDD 转换操作

（1）reduceByKey(func)

reduceByKey(func)的功能是使用 func 函数合并具有相同键的值。实现根据 RDD 中的相同键的值进行聚合相加功能的示例代码如下。

Python 示例代码如下。

```
>>> pairRDD = sc.parallelize([("Hadoop",1),("Spark",1),("Hive",1),
          ("Spark",1)])
>>> pairRDD.reduceByKey(lambda a,b:a+b).foreach(print)
('Spark', 2)
('Hive', 1)
('Hadoop', 1)
```

Java 示例代码如下。

```
import org.apache.spark.SparkConf;
import org.apache.spark.api.java.JavaPairRDD;
import org.apache.spark.api.java.JavaSparkContext;
import scala.Tuple2;
import java.util.Arrays;

public class ReduceByKeyJava {
    public static void main(String[] args) {
        // 初始化 Spark 配置和上下文
        SparkConf conf = new SparkConf().setAppName(
                "ReduceByKeyJavaApp").setMaster("local");
        JavaSparkContext sc = new JavaSparkContext(conf);

        // 创建一个键值对 RDD
        JavaPairRDD<String, Integer> pairRDD = sc.parallelizePairs(
            Arrays.asList(
            new Tuple2<>("Hadoop", 1),
            new Tuple2<>("Spark", 1),
            new Tuple2<>("Hive", 1),
            new Tuple2<>("Spark", 1)
        ));

        // 使用 reduceByKey()函数合并相同键的值
        JavaPairRDD<String, Integer> reducedRDD = pairRDD.reduceByKey((a, b) ->
          a + b);

        // 打印结果
        reducedRDD.foreach(tuple -> System.out.println("(" + tuple._1() + "," +
            tuple._2() + ")"));

        // 关闭 Spark 上下文
        sc.close();
    }
}
```

（2）groupByKey()

groupByKey()的功能是对具有相同键的值进行分组。

例如，对 4 个键值对("spark", 1)、("spark", 2)、("hadoop", 3)和("hadoop", 5)，采用 groupByKey()后得到的结果是("spark", (1, 2))和("hadoop", (3, 5))。

Python 示例代码如下。

```
>>> list = [("spark",1),("spark",2),("hadoop",3),("hadoop",5)]
```

```
>>> pairRDD = sc.parallelize(list)
>>> pairRDD.groupByKey()
PythonRDD[27] at RDD at PythonRDD.scala:48
>>> pairRDD.groupByKey().foreach(print)
('hadoop', <pyspark.resultiterable.ResultIterable object at 0x7f2c1093ecf8>)
('spark', <pyspark.resultiterable.ResultIterable object at 0x7f2c1093ecf8>)
```

Java 示例代码如下。

```
import org.apache.spark.SparkConf;
import org.apache.spark.api.java.JavaPairRDD;
import org.apache.spark.api.java.JavaSparkContext;
import scala.Tuple2;
import java.util.Arrays;
import java.util.List;

public class GroupByKeyJava {
    public static void main(String[] args) {
        // 初始化 Spark 配置和上下文
        SparkConf conf = new SparkConf().setAppName(
                        "GroupByKeyJavaApp").setMaster("local");
        JavaSparkContext sc = new JavaSparkContext(conf);

        // 创建一个键值对 RDD
        JavaPairRDD<String, Integer> pairRDD = sc.parallelizePairs(
        Arrays.asList(
            new Tuple2<>("spark", 1),
            new Tuple2<>("spark", 2),
            new Tuple2<>("hadoop", 3),
            new Tuple2<>("hadoop", 5)
        ));

        // 使用 groupByKey()函数
        JavaPairRDD<String, Iterable<Integer>> groupedRDD =
        pairRDD.groupByKey();

        // 打印结果
        groupedRDD.foreach(tuple -> {
            List<Integer> valuesList = Arrays.asList(
                                    tuple._2().toArray(new Integer[0]));
            System.out.println("(" + tuple._1() + ", " + valuesList + ")");
        });

        // 关闭 Spark 上下文
        sc.close();
    }
}
```

（3）reduceByKey()和 groupByKey()的区别。

reduceByKey()用于对每个键对应的多个值进行合并操作，能够在本地先进行合并操作。合并操作可以通过函数进行自定义。

groupByKey()也是对每个键进行操作，但只生成一个队列。groupByKey()本身不能自定义函数，需要先生成 RDD，然后才能对此 RDD 通过 map()函数进行自定义函数操作。

比较 reduceByKey()和 groupByKey()，实现词频统计功能的示例代码如下。

Python 示例代码如下。

```
>>> words = ["one", "two", "two", "three", "three", "three"]
>>> wordPairsRDD = sc.parallelize(words).map(lambda word:(word, 1))
>>> wordCountsWithReduce = wordPairsRDD.reduceByKey(lambda a,b:a+b)
>>> wordCountsWithReduce.foreach(print)
('one', 1)
('two', 2)
('three', 3)
>>> wordCountsWithGroup = wordPairsRDD.groupByKey(). \
... map(lambda t:(t[0],sum(t[1])))
>>> wordCountsWithGroup.foreach(print)
('two', 2)
('three', 3)
('one', 1)
```

Java 示例代码如下。

```
import org.apache.spark.SparkConf;
import org.apache.spark.api.java.JavaPairRDD;
import org.apache.spark.api.java.JavaRDD;
import org.apache.spark.api.java.JavaSparkContext;
import scala.Tuple2;
import java.util.Arrays;
import java.util.List;

public class WordCountJava {
    public static void main(String[] args) {
        // 初始化 Spark 配置和上下文
        SparkConf conf = new SparkConf().setAppName(
                    "WordCountJavaApp").setMaster("local");
        JavaSparkContext sc = new JavaSparkContext(conf);

        // 创建一个 Java RDD
        List<String> words = Arrays.asList("one", "two", "two", "three",
                          "three", "three");
        JavaRDD<String> wordsRDD = sc.parallelize(words);

        // 转换成键值对 RDD 并使用 reduceByKey()方法
        JavaPairRDD<String, Integer> wordPairsRDD =
        wordsRDD.mapToPair(word -> new Tuple2<>(word, 1));
        JavaPairRDD<String, Integer> wordCountsWithReduce =
        wordPairsRDD.reduceByKey((a, b) -> a + b);
        wordCountsWithReduce.foreach(tuple -> System.out.println("
        (" + tuple._1() + ", " + tuple._2() + ")"));

        // 使用 groupByKey()方法
```

```
        JavaPairRDD<String, Integer> wordCountsWithGroup =
        wordPairsRDD.groupByKey().mapToPair(tuple -> new Tuple2<>(tuple._1(),
        sum(tuple._2())));
        wordCountsWithGroup.foreach(tuple -> System.out.println("
        (" + tuple._1() + ", " + tuple._2() + ")"));

        // 关闭 Spark 上下文
        sc.close();
    }

    // 辅助函数，用于计算 Iterable 内所有元素的总和
    private static int sum(Iterable<Integer> nums) {
        int sum = 0;
        for (int num : nums) {
            sum += num;
        }
        return sum;
    }
}
```

上面两段代码得到的 wordCountsWithReduce 是完全一样的，但是，它们的内部运算过程是不同的。

7.2.3　数据读/写操作

1．本地文件系统的数据读/写

（1）从文件中读取数据创建 RDD

从文件中读取数据，并创建 RDD 的示例代码如下。

Python 示例代码如下。

```
>>> textFile = sc.\
... textFile("file:///usr/local/spark/mycode/rdd/word.txt")
>>> textFile.first()
'Hadoop is good'
```

Java 示例代码如下。

```
JavaRDD<String> textFile =
sc.textFile("file:///usr/local/spark/mycode/rdd/word.txt");
System.out.println(textFile.first());
```

因为 Spark 采用了惰性机制，在执行转换操作时，即使输入了错误的语句，Spark Shell 也不会马上报错（假设 word123.txt 文件不存在）。

Python 示例代码如下。

```
>>> textFile = sc.\
... textFile("file:///usr/local/spark/mycode/wordcount/word123.txt")
```

Java 示例代码如下。

```
JavaRDD<String> textFile =
sc.textFile("file:///usr/local/spark/mycode/rdd/word123.txt");
```

（2）把 RDD 写入文本文件。

把 RDD 写入文本文件的示例代码如下。

Python 示例代码如下。

```
>>> textFile = sc.\
... textFile("file:///usr/local/spark/mycode/rdd/word.txt")
>>> textFile.\
... saveAsTextFile("file:///usr/local/spark/mycode/rdd/writeback")
```

Java 示例代码如下。

```
import org.apache.spark.SparkConf;
import org.apache.spark.api.java.JavaRDD;
import org.apache.spark.api.java.JavaSparkContext;

public class SaveTextFileJava {
    public static void main(String[] args) {
        // 初始化 Spark 配置和上下文
        SparkConf conf =
        new SparkConf().setAppName("SaveTextFileJavaApp").setMaster("local");
        JavaSparkContext sc = new JavaSparkContext(conf);

        // 从文件中读取数据创建 RDD
        JavaRDD<String> textFile =
        sc.textFile("file:///usr/local/spark/mycode/rdd/word.txt");

        // 把 RDD 写入文本文件中
        textFile.saveAsTextFile(
        "file:///usr/local/spark/mycode/rdd/writeback");

        // 关闭 Spark 上下文
        sc.close();
    }
}
```

如果想再次把数据加载在 RDD 中，只需要使用 writeback 这个目录即可，具体示例代码如下。

Python 示例代码如下。

```
>>> textFile = sc.\
... textFile("file:///usr/local/spark/mycode/rdd/writeback")
```

Java 示例代码如下。

```
import org.apache.spark.SparkConf;
import org.apache.spark.api.java.JavaRDD;
import org.apache.spark.api.java.JavaSparkContext;

public class ReadTextFileJava {
    public static void main(String[] args) {
        // 初始化 Spark 配置和上下文
        SparkConf conf =
        new SparkConf().setAppName("ReadTextFileJavaApp").setMaster("local");
```

```
        JavaSparkContext sc = new JavaSparkContext(conf);

        // 从文本文件读入数据到 RDD
        JavaRDD<String> textFile =
        sc.textFile("file:///usr/local/spark/mycode/rdd/writeback");

        // 可选操作：打印 RDD 的内容（例如，第一行）
        System.out.println(textFile.first());

        // 关闭 Spark 上下文
        sc.close();
    }
}
```

2. HDFS 的数据读/写

从 HDFS 中读取数据可以采用 textFile()方法，我们可以为该方法提供一个 HDFS 文件或目录地址。如果提供的是一个文件地址，textFile()方法会加载该文件。如果提供的是一个目录地址，textFile()方法会加载该目录下的所有文件的数据。

从 HDFS 中读取数据的示例代码如下（HDFS 文件）。

Python 示例代码如下。

```
>>> textFile = sc.textFile("hdfs:// localhost:9000/user/hadoop/word.txt")
>>> textFile.first()
```

Java 示例代码如下。

```
import org.apache.spark.SparkConf;
import org.apache.spark.api.java.JavaRDD;
import org.apache.spark.api.java.JavaSparkContext;

public class ReadHDFSTextFileJava {
    public static void main(String[] args) {
        // 初始化 Spark 配置和上下文
        SparkConf conf =
        new SparkConf().setAppName("ReadHDFSTextFileJavaApp");
        JavaSparkContext sc = new JavaSparkContext(conf);

        // 从 HDFS 中的文本文件读入数据到 RDD
        JavaRDD<String> textFile =
        sc.textFile("hdfs:// localhost:9000/user/hadoop/word.txt");

        // 打印 RDD 的第一行
        System.out.println(textFile.first());

        // 关闭 Spark 上下文
        sc.close();
    }
}
```

从 HDFS 中读取数据的示例代码如下（目录地址或文件地址）。

Python 示例示例如下，这 3 条语句都是等价的。

```
>>> textFile = sc.textFile("hdfs:// localhost:9000/user/hadoop/word.txt")
>>> textFile = sc.textFile("/user/hadoop/word.txt")
>>> textFile = sc.textFile("word.txt")
```

Java 示例代码如下。

```
import org.apache.spark.SparkConf;
import org.apache.spark.api.java.JavaRDD;
import org.apache.spark.api.java.JavaSparkContext;

public class ReadHDFSEquivalentJava {
    public static void main(String[] args) {
        // 初始化 Spark 配置和上下文
        SparkConf conf =
        new SparkConf().setAppName("ReadHDFSEquivalentJavaApp");
        JavaSparkContext sc = new JavaSparkContext(conf);

        // 从 HDFS 完整路径加载文本文件到 RDD
        JavaRDD<String> textFile1 =
        sc.textFile("hdfs:// localhost:9000/user/hadoop/word.txt");

        // 从 HDFS 根目录的相对路径加载文本文件到 RDD
        JavaRDD<String> textFile2 = sc.textFile("/user/hadoop/word.txt");

        // 使用文件名（这需要该文件在 Spark 的默认文件系统上，通常是 HDFS 的根目录）
        JavaRDD<String> textFile3 = sc.textFile("word.txt");

        // 打印 3 个 RDD 的第一行，以验证它们是否相同
        System.out.println(textFile1.first());
        System.out.println(textFile2.first());
        System.out.println(textFile3.first());

        // 关闭 Spark 上下文
        sc.close();
    }
}
```

同样，我们可以使用 saveAsTextFile()方法把 RDD 中的数据写入到 HDFS 文件中，示例代码如下。

Python 示例代码如下。

```
>>> textFile = sc.textFile("word.txt")
>>> textFile.saveAsTextFile("writeback")
```

Java 示例代码如下。

```
import org.apache.spark.SparkConf;
import org.apache.spark.api.java.JavaRDD;
import org.apache.spark.api.java.JavaSparkContext;

public class ReadAndWriteHDFSJava {
    public static void main(String[] args) {
        // 初始化 Spark 配置和上下文
        SparkConf conf = new SparkConf().setAppName("ReadAndWriteHDFSJavaApp");
```

```
        JavaSparkContext sc = new JavaSparkContext(conf);

        // 从默认文件系统的相对路径加载文本文件到 RDD
        JavaRDD<String> textFile = sc.textFile("word.txt");

        // 保存 RDD 到默认文件系统的相对路径
        textFile.saveAsTextFile("writeback");

        // 关闭 Spark 上下文
        sc.close();
    }
}
```

7.3 编程实现

7.3.1 任务 1：取出排名前五的订单支付金额

文件 file1.txt、file2.txt 中有字段 orderid、userid、payment、productid，现在要求把两个文件合并，并求排名前五的 payment 值。两个文件的具体内容如下。

```
  file1.txt                        file2.txt
1,1768,50,155                   100,4287,226,233
2,1218, 600,211                 101,6562,489,124
3,2239,788,242                  102,1124,33,17
4,3101,28,599                   103,3267,159,179
5,4899,290,129                  104,4569,57,125
6,3110,54,1201                  105,1438,37,116
7,4436,259,877
8,2369,7890,27
```

实现针对 2 份购物记录数据，进行合并统计，求得购物金额最多的前五名功能的完整示例代码如下。

Python 示例代码如下。

```
# !/usr/bin/env python3
from pyspark import SparkConf, SparkContext

conf = SparkConf().setMaster("local").setAppName("ReadHBase")
sc = SparkContext(conf = conf)
lines = sc.textFile("file:///usr/local/spark/mycode/rdd/file*.txt")
result1 = lines.filter(lambda line:(len(line.strip()) > 0) and
        (len(line.split(","))== 4))
result2 = result1.map(lambda x:x.split(",")[2])
result3 = result2.map(lambda x:(int(x),""))
result4 = result3.repartition(1)
result5 = result4.sortByKey(False)
result6 = result5.map(lambda x:x[0])
```

```
result7 = result6.take(5)
for a in result7:
    print(a)
```

Java 示例代码如下。

```
import org.apache.spark.SparkConf;
import org.apache.spark.api.java.JavaPairRDD;
import org.apache.spark.api.java.JavaRDD;
import org.apache.spark.api.java.JavaSparkContext;
import scala.Tuple2;
import java.util.Arrays;
import java.util.List;

public class TopNJava {
    public static void main(String[] args) {
        SparkConf conf = new SparkConf().setAppName(
                        "TopNJavaApp").setMaster("local");
        JavaSparkContext sc = new JavaSparkContext(conf);

        JavaRDD<String> lines = sc.textFile(
                            "file:///usr/local/spark/mycode/rdd/file*.txt");

        JavaRDD<String> result1 = lines.filter(line -> {
            String[] parts = line.split(",");
            return !line.trim().isEmpty() && parts.length == 4;
        });

        JavaRDD<String> result2 = result1.map(line -> line.split(",")[2]);

        JavaPairRDD<Integer, String> result3 = result2.mapToPair(payment -> {
            int paymentValue = Integer.parseInt(payment);
            return new Tuple2<>(paymentValue, "");
        });

        JavaPairRDD<Integer, String> result4 = result3.repartition(1);
        JavaPairRDD<Integer, String> result5 = result4.sortByKey(false);

        List<Integer> result7 = result5.keys().take(5);

        for (int a : result7) {
            System.out.println(a);
        }

        sc.close();
    }
}
```

代码解读如下。

实现文件读取为 RDD 功能的示例代码如下。

Python 示例代码如下。

```
lines = sc.textFile("file:///usr/local/spark/mycode/rdd/file*.txt")
```

Java 示例代码如下。

```
JavaRDD<String> lines = sc.textFile("file:///usr/local/spark/mycode/rdd/file*.txt");
```

上述代码的执行效果如图 7-13 所示。

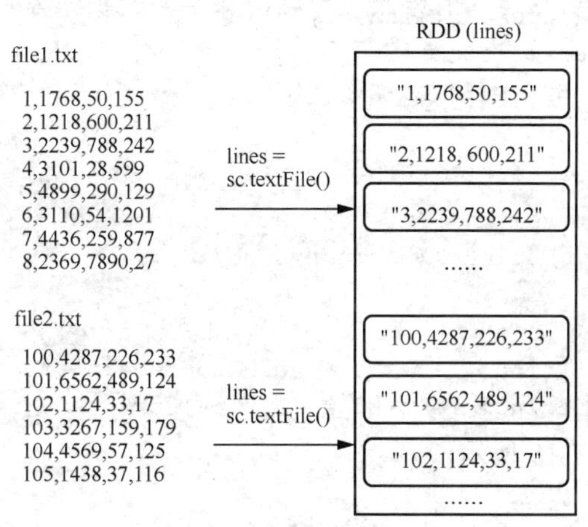

图 7-13　文件读取为 RDD

读取 payment 数据，示例代码如下。

Python 示例代码如下。

```
result1 = lines.filter(lambda line:(len(line.strip()) > 0) and (len(line.split
(","))== 4))
result2 = result1.map(lambda x:x.split(",")[2])
```

Java 示例代码如下。

```
JavaRDD<String> result1 = lines.filter(line -> {
    String[] parts = line.split(",");
    return !line.trim().isEmpty() && parts.length == 4;
       });

JavaRDD<String> result2 = result1.map(line -> line.split(",")[2]);
```

代码执行效果如图 7-14 所示。

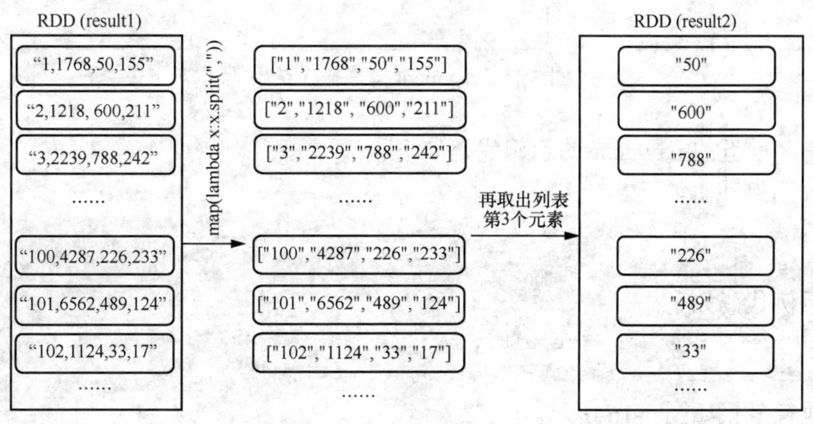

图 7-14　读取 payment

实现 payment 转换为（数值类型的 payment，""）的形式，重新化为一个分区，最后按照 key 进行降序排序功能的示例代码如下。

Python 示例代码如下。

```
result3 = result2.map(lambda x:(int(x),""))
result4 = result3.repartition(1)
result5 = result4.sortByKey(false)
```

Java 示例代码如下。

```
JavaPairRDD<Integer, String> result3 = result2.mapToPair(payment -> {
            int paymentValue = Integer.parseInt(payment);
            return new Tuple2<>(paymentValue, "");
        });

JavaPairRDD<Integer, String> result4 = result3.repartition(1);
JavaPairRDD<Integer, String> result5 = result4.sortByKey(false);
```

代码执行效果如图 7-15 所示。

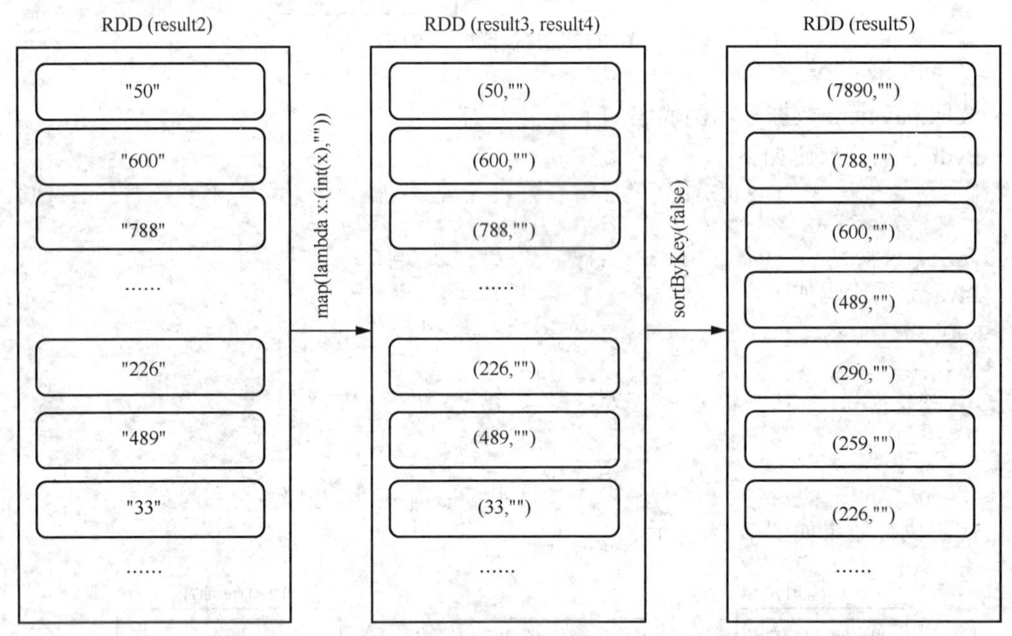

图 7-15 payment 降序排序结果

获取排名前五的 payment 的示例代码如下。

Python 示例代码如下。

```
result6 = result5.map(lambda x:x[0])
result7 = result6.take(5)
```

Java 示例代码如下。

```
JavaRDD<Integer> result6 = result5.keys();
List<Integer> result7 = result6.take(5);
```

代码执行效果如图 7-16 所示。

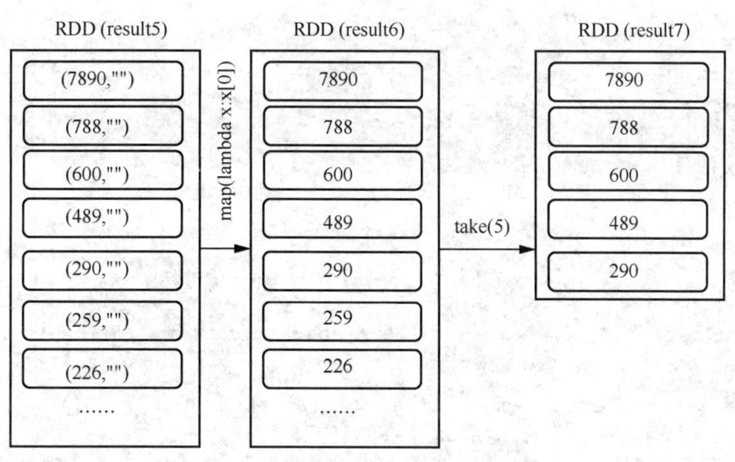

图 7-16　获取排名前五的 payment

7.3.2　任务 2：文件排序

有多个输入文件，每个文件中的每行内容均为一个整数。要求读取所有文件中的整数并进行排序，排序后的结果输出到一个新的文件中。在这个新文件中，每行有两个整数，第一个整数为第二个整数的排序位次，第二个整数为排序的整数。

输入文件的内容如下。

```
file1.txt:            file2.txt              file3.txt
33                      4                      1
37                      16                     45
12                      39                     25
40                      5
```

输出文件的内容如下。

```
1   1
2   4
3   5
4   12
5   16
6   25
7   33
8   37
9   39
10  40
11  45
```

实现文件排序功能的完整示例代码如下。

Python 示例代码如下（FileSort.py）。

```
# !/usr/bin/env python3

from pyspark import SparkConf, SparkContext
```

```
index = 0
def getindex():
    global index
    index+=1
    return index

def main():
    conf = SparkConf().setMaster("local[1]").setAppName("FileSort")
    sc = SparkContext(conf = conf)
    lines = sc.textFile("file:///usr/local/spark/mycode/rdd/filesort/file*.txt")
    index = 0
    result1 = lines.filter(lambda line:(len(line.strip()) > 0))
    result2 = result1.map(lambda x:(int(x.strip()),""))
    result3 = result2.repartition(1)
    result4 = result3.sortByKey(True)
    result5 = result4.map(lambda x:x[0])
    result6 = result5.map(lambda x:(getindex(),x))
    result6.foreach(print)
    result6.saveAsTextFile(
    "file:// /usr/local/spark/mycode/rdd/filesort/sortresult")
if __name__ == '__main__':
    main()
```

Java 示例代码如下。

```
import org.apache.spark.SparkConf;
import org.apache.spark.api.java.JavaPairRDD;
import org.apache.spark.api.java.JavaRDD;
import org.apache.spark.api.java.JavaSparkContext;
import scala.Tuple2;

import java.util.List;

public class FileSortJava {
    public static void main(String[] args) {
        SparkConf conf =
        new SparkConf().setAppName("FileSortJavaApp").setMaster("local[1]");
        JavaSparkContext sc = new JavaSparkContext(conf);

        JavaRDD<String> lines =
        sc.textFile("file:///usr/local/spark/mycode/rdd/filesort/file*.txt");

        JavaRDD<String> result1 = lines.filter(line -> !line.trim().isEmpty());
        JavaPairRDD<Integer, String> result2 = result1.mapToPair(line -> {
            int value = Integer.parseInt(line.trim());
            return new Tuple2<>(value, "");
        });

        JavaPairRDD<Integer, String> result3 = result2.repartition(1);
```

```
        JavaPairRDD<Integer, String> result4 = result3.sortByKey(true);
        JavaRDD<Integer> result5 = result4.keys();

        JavaPairRDD<Integer, Integer> result6 =
        result5.zipWithIndex().mapToPair(t -> {
            int index = (int) t._2 + 1;  // 1-based index
            int value = t._1;
            return new Tuple2<>(index, value);
        });

        result6.foreach(tuple -> System.out.println(tuple._1() + "\t" + tuple._2()));
        result6.saveAsTextFile("
        file:///usr/local/spark/mycode/rdd/filesort/sortresult");

        sc.close();
    }
}
```

代码解释如下。

实现读取文件内容，将内容转换为 RDD，同时过滤空白行功能的示例代码如下。

Python 示例代码如下。

```
lines = sc.textFile("file:///usr/local/spark/mycode/rdd/filesort/file*.txt")
result1 = lines.filter(lambda line:(len(line.strip()) > 0))
```

Java 示例代码如下。

```
JavaRDD<String> lines = sc.textFile("file:///usr/local/spark/mycode/rdd/filesort/
file*.txt");

JavaRDD<String> result1 = lines.filter(line -> !line.trim().isEmpty());
```

代码执行效果如图 7-17 所示。

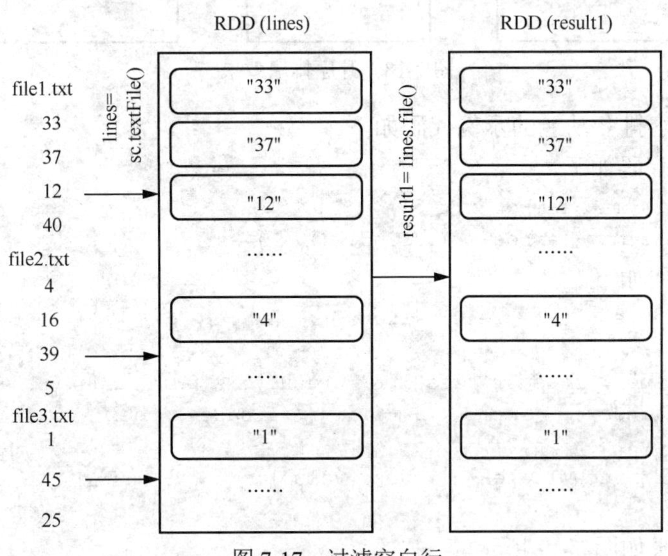

图 7-17　过滤空白行

实现文本内容转换为（数值类型的文本内容，""）的形式，重新化为一个分区，最后按照 key 进行升序排序功能的示例代码如下。

Python 示例代码如下。

```
result2 = result1.map(lambda x:(int(x.strip()),""))
result3 = result2.repartition(1)
result4 = result3.sortByKey(true)
```

Java 示例代码如下。

```
JavaPairRDD<Integer, String> result2 = result1.mapToPair(line -> {
            int value = Integer.parseInt(line.trim());
            return new Tuple2<>(value, "");
        });

JavaPairRDD<Integer, String> result3 = result2.repartition(1);
JavaPairRDD<Integer, String> result4 = result3.sortByKey(true);
```

代码执行效果如图 7-18 所示。

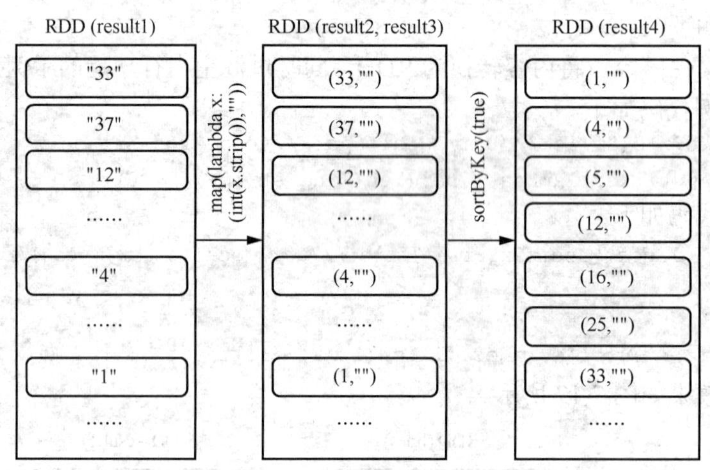

图 7-18　升序排序效果

实现文本内容排名功能的示例代码如下。

Python 示例代码如下。

```
result5 = result4.map(lambda x:x[0])
result6 = result5.map(lambda x:(getindex(),x))
```

Java 示例代码如下。

```
JavaRDD<Integer> result5 = result4.keys();

JavaPairRDD<Integer, Integer> result6 = result5.zipWithIndex().mapToPair(t -> {
            int index = (int) t._2 + 1;  // 1-based index
            int value = t._1;
            return new Tuple2<>(index, value);
        });
```

代码执行效果如图 7-19 所示。

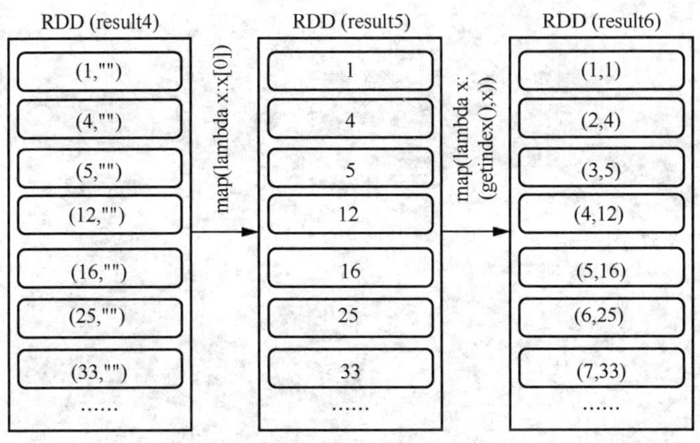

图 7-19 文本内容排名

实现排名后内容写入文件功能的示例代码如下。

Python 示例代码如下。

```
result6.saveAsTextFile("file:///usr/local/spark/mycode/rdd/filesort/sortresult")
```

Java 示例代码如下。

```
result6.saveAsTextFile("file:///usr/local/spark/mycode/rdd/filesort/sortresult");
```

代码执行效果如图 7-20 所示。

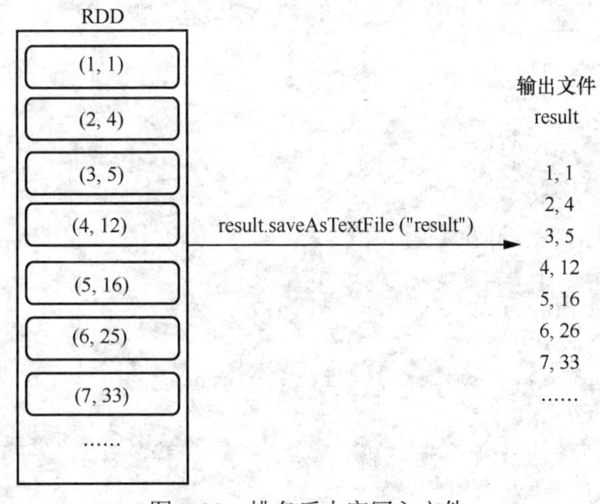

图 7-20 排名后内容写入文件

7.3.3 任务 3：二次排序

对于一个给定的文件（数据如下面的 file1.txt 文件所示），请对数据进行排序，首先根据第 1 列数据降序排序，如果第 1 列数据相等，则根据第 2 列数据降序排序。

输入文件的内容如下。

```
file1.txt
5   3
```

```
1   6
4   9
8   3
4   7
5   6
3   2
```

输出结果的内容如下。

```
8    3
5    6
5    3
4    9
4    7
3    2
1    6
```

实现二次排序功能的完整示例代码如下。

Python 示例代码如下。

SecondarySortApp.py 代码如下。

```
# !/usr/bin/env python3

from operator import gt
from pyspark import SparkContext, SparkConf

class SecondarySortKey():
    def __init__(self, k):
        self.column1 = k[0]
        self.column2 = k[1]

    def __gt__(self, other):
        if other.column1 == self.column1:
            return gt(self.column2,other.column2)
        else:
            return gt(self.column1, other.column1)
def main():
    conf = SparkConf().setAppName('spark_sort').setMaster('local[1]')
    sc = SparkContext(conf=conf)
    file="file:///usr/local/spark/mycode/rdd/secondarysort/file1.txt"
    rdd1 = sc.textFile(file)
    rdd2 = rdd1.filter(lambda x:(len(x.strip()) > 0))
    rdd3 = rdd2.map(lambda x:((int(x.split(" ")[0]),int(x.split(" ")[1])),x))
    rdd4 = rdd3.map(lambda x: (SecondarySortKey(x[0]),x[1]))
    rdd5 = rdd4.sortByKey(False)
    rdd6 = rdd5.map(lambda x:x[1])
    rdd6.foreach(print)

if __name__ == '__main__':
    main()
```

Java 示例代码如下。

```
import org.apache.spark.SparkConf;
import org.apache.spark.api.java.JavaPairRDD;
import org.apache.spark.api.java.JavaRDD;
import org.apache.spark.api.java.JavaSparkContext;
import scala.Tuple2;
import java.io.Serializable;
import java.util.Comparator;

// 自定义的排序键
class SecondarySortKey implements Serializable, Comparable<SecondarySortKey> {
    private int column1;
    private int column2;

    public SecondarySortKey(int column1, int column2) {
        this.column1 = column1;
        this.column2 = column2;
    }

    @Override
    public int compareTo(SecondarySortKey other) {
        if (this.column1 == other.column1) {
            return Integer.compare(this.column2, other.column2);
        } else {
            return Integer.compare(this.column1, other.column1);
        }
    }
}

public class SecondarySortApp {
    public static void main(String[] args) {
        SparkConf conf = new SparkConf().setAppName(
                "SecondarySortApp").setMaster("local[1]");
        JavaSparkContext sc = new JavaSparkContext(conf);

        String file = "file:///usr/local/spark/mycode/rdd/secondarysort/file1.txt";
        JavaRDD<String> rdd1 = sc.textFile(file);

        JavaPairRDD<Tuple2<Integer, Integer>, String> rdd3 = rdd1
            .filter(line -> !line.trim().isEmpty())
            .mapToPair(line -> {
                String[] parts = line.split(" ");
                int key1 = Integer.parseInt(parts[0]);
                int key2 = Integer.parseInt(parts[1]);
                Tuple2<Integer, Integer> key = new Tuple2<>(key1, key2);
                return new Tuple2<>(key, line);
            });
```

```
        JavaPairRDD<SecondarySortKey, String> rdd4 = rdd3.mapToPair(tuple -> {
            SecondarySortKey key = new SecondarySortKey(tuple._1._1, tuple._1._2);
            return new Tuple2<>(key, tuple._2);
        });

        JavaPairRDD<SecondarySortKey, String> rdd5 = rdd4.sortByKey();

        JavaRDD<String> rdd6 = rdd5.map(Tuple2::_2);

        rdd6.foreach(System.out::println);

        sc.close();
    }
}
```

代码解释如下。

实现过滤空白行功能的示例代码如下。

Python 示例代码如下。

```
rdd1 = sc.textFile(file)
rdd2 = rdd1.filter(lambda x:(len(x.strip()) > 0))
```

Java 示例代码如下。

```
JavaRDD<String> rdd1 = sc.textFile(file);

JavaRDD<String> rdd2 = rdd1.filter(line -> !line.trim().isEmpty());
```

代码执行效果如图 7-21 所示。

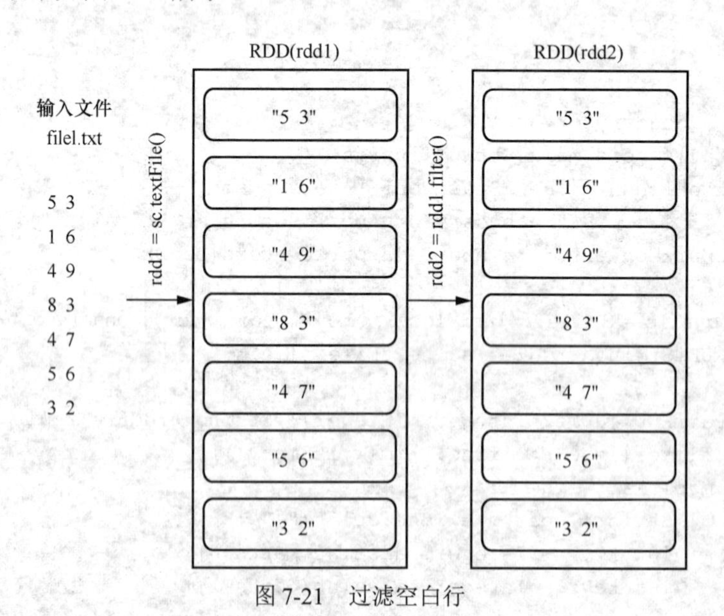

图 7-21　过滤空白行

二次排序示例代码如下。

Python 示例代码如下。

```
rdd3 = rdd2.map(lambda x:((int(x.split(" ")[0]),int(x.split(" ")[1])),x))
rdd4 = rdd3.map(lambda x: (SecondarySortKey(x[0]),x[1]))
```

Java 示例代码如下。

```
JavaPairRDD<Tuple2<Integer, Integer>, String> rdd3 = rdd1
        .filter(line -> !line.trim().isEmpty())
        .mapToPair(line -> {
            String[] parts = line.split(" ");
            int key1 = Integer.parseInt(parts[0]);
            int key2 = Integer.parseInt(parts[1]);
            Tuple2<Integer, Integer> key = new Tuple2<>(key1, key2);
            return new Tuple2<>(key, line);
        });

JavaPairRDD<SecondarySortKey, String> rdd4 = rdd3.mapToPair(tuple -> {
        SecondarySortKey key = new SecondarySortKey(tuple._1._1, tuple._1._2);
        return new Tuple2<>(key, tuple._2);
    });
```

代码执行效果如图 7-22 所示。

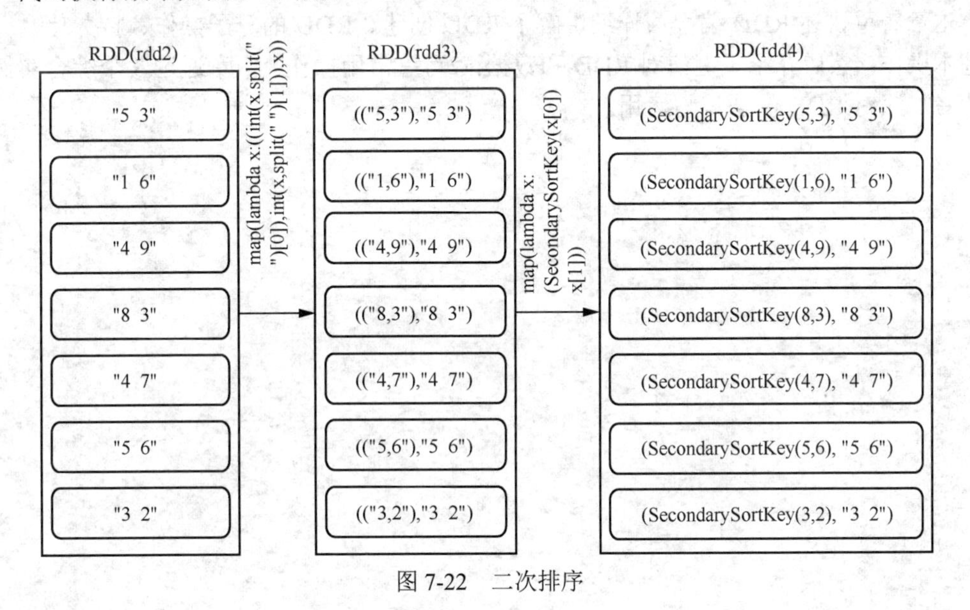

图 7-22　二次排序

实现二次排序后，按照原始格式输出的示例代码如下。

Python 示例代码如下。

```
rdd5 = rdd4.sortByKey(False)
rdd6 = rdd5.map(lambda x:x[1])
```

Java 示例代码如下。

```
JavaPairRDD<SecondarySortKey, String> rdd5 = rdd4.sortByKey();
JavaRDD<String> rdd6 = rdd5.map(Tuple2::_2);
```

代码执行效果如图 7-23 所示。

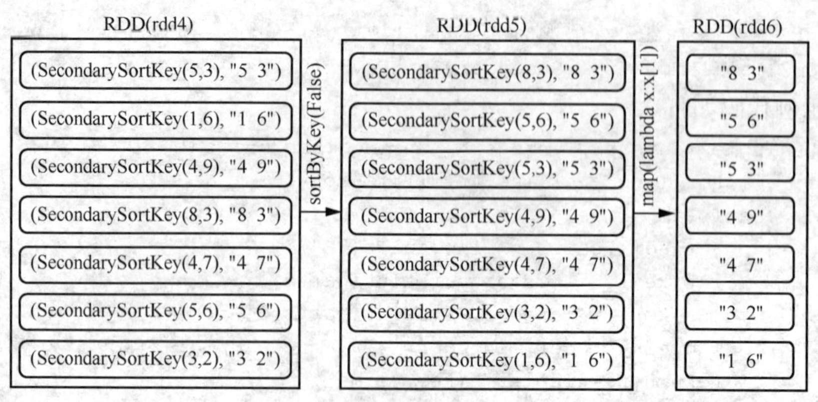

图 7-23　原始格式输出

7.4 本章小结

本章着重讲述 RDD 编程，详细讲解了 RDD 创建、RDD 的算子操作、持久化、分区等基本操作；然后介绍了键值对 RDD 和数据读/写等常用操作。本章通过 3 个综合实例帮助读者加深对 RDD 的认识与运用。

第 8 章 Spark SQL：结构化数据处理

随着 Spark 版本的不断更新，Spark RDD 的局限性逐渐显现：Spark RDD 是一个底层组件，在实际开发中的效率并不高。为了解决这个问题，Spark 的开发者对 Spark RDD 进行了高层次的封装，从而推出了 Spark DataFrame 和 Spark SQL。这两个新组件的出现使 Spark 的程序开发变得与编写常规单机程序一样简洁，但也导致 Spark RDD 逐步失去主导地位。

【学习目标】

1．了解 Spark SQL 的基本概念和特征原理。

2．学会如何创建 DataFrame 对象及如何操作 DataFrame 对象。

3．认识 Spark SQL 与 MySQL 的交互。

8.1 Spark SQL 概述

【任务描述】了解 Spark SQL 的基本概念，对 Spark SQL 相关环境进行配置，掌握 Spark SQL 与 Shell 的交互方式。

8.1.1 Spark SQL 简介

从 Spark 1.3 版本起，基于原有的 SchemaRDD，Spark SQL 引入了与 R 语和 Pandas 相似的 DataFrame API。这个新的 API 不仅大大简化了普通用户的学习过程，而且支持 Scala、Java 和 Python 3 种主流编程语言。更重要的是，由于 DataFrame 是基于 SchemaRDD 构建的，因此它更适合应对分布式大数据处理的挑战。

在 Spark 的生态系统中，DataFrame 是以 RDD 为底层基础构建的，是一个分布式的数据集，与传统数据库的二维表非常相似。RDD、DataFrame 的结构如图 8-1 所示。

DataFrame 与 RDD 的显著区别在于，DataFrame 携带了 Schema 元信息，即它代表的二维数据集明确了每一列的名称与数据类型。这种明确的结构化信息使 Spark SQL 能够更深入地“理解”数据结构，从而对 DataFrame 背后的数据及其上的转换操作进行更加精准的优化，大大提高了执行效率。

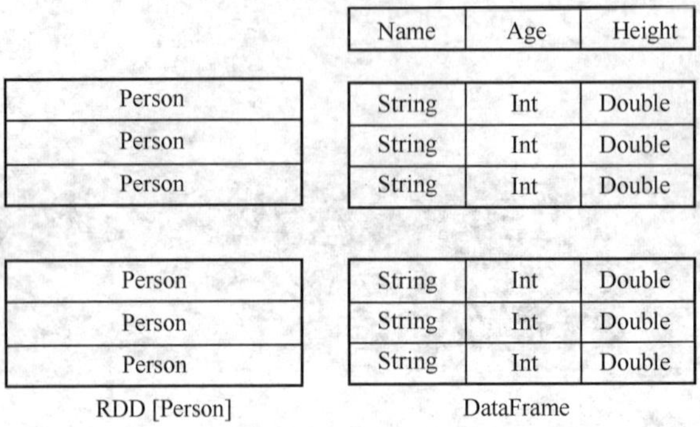

Person
Person
Person
Person
Person
Person

RDD [Person]

Name	Age	Height
String	Int	Double
String	Int	Double
String	Int	Double
String	Int	Double
String	Int	Double
String	Int	Double

DataFrame

图 8-1　RDD、DataFrame 的结构

尽管关系数据库已经深入人心并被广泛使用，但在大数据时代，关系数据库已不能完全满足日益增长的需求。

① 用户希望能够对来自不同数据源的数据进行操作，这些数据可能是结构化的、半结构化的或非结构化的。

② 用户希望执行更高级的数据分析，例如机器学习和图像处理。

实际的大数据应用场景中经常需要将关系型查询与复杂的分析算法（例如机器学习或图像处理）结合起来。但遗憾的是，市场上缺乏能够满足这种需求的系统。

此时，Spark SQL 便发挥了作用。

① 它提供了 DataFrame API，使用户可以对各种内部和外部数据源进行关系型操作。

② 它支持大数据领域的众多数据源和数据分析算法。

总而言之，Spark SQL 实现了关系数据库的结构化数据管理与机器学习算法的数据处理之间的完美融合。

8.1.2　Spark SQL CLI 配置

Spark SQL 与 Hive 具有良好的兼容性，因此 Spark SQL 能够支持 Hive 的表访问、用户自定义函数（User Define Function，UDF）和 Hive 查询语言（HiveQL）。从 Spark 1.1 版本开始，Spark 引入了 Spark SQL CLI 和 ThriftServer，这使得熟悉 Hive 和习惯使用命令行的关系数据库管理系统管理员能够更轻松地使用 Spark SQL。

若想通过 Spark SQL CLI 来访问和操作 Hive 表数据，需要按照以下步骤配置 Spark SQL 的环境（前提是 Spark 集群已经配置完毕），以便将 Spark SQL 与一个已部署的 Hive 实例连接起来。需要注意的是，即使没有 Hive 的部署实例，Spark SQL 也可以独立运行。在这种情况下，Spark SQL 会在当前的工作目录下创建一个名为 metastore_db 的 Hive 元数据库。接下来，我们将介绍如何配置 Spark SQL，以使用 Hive 环境。

① 将 hive-site.xml 复制到/usr/local/spark/conf 目录下，代码如下。

```
$cd /usr/local/hive/conf/hive-site.xml /usr/local/spark/conf/
```

② 在/usr/local/spark/conf/spark-env.sh 文件中配置 MySQL 驱动。执行命令 vim /usr/local/spark/conf/spark-env.sh 来打开文件，在文件末尾添加以下内容。

```
$vim conf/spark-env.sh

export SPARK_CLASSPATH=$HIVE_HOME/lib/mysql-connector.java-5.1.32-bin.jar3
```

③ 启动 MySQL 服务，执行以下命令。

```
$service mysqld start
```

④ 启动 Hive 的 metastore 服务，执行以下命令。

```
$hive --service metastore &
```

⑤ 修改日志级别。执行 cd/usr/local/spark/conf/命令进入 Spark 的 conf 目录，将该目录下的 log4j.properties.template 文件复制为 log4j.properties，执行 vimlog4j.properties 命令来打开文件，修改文件中“log4j.rootCategory”的值，修改后的内容如下。

```
#Set everything to be logged to the console
log4j.rootCategory = WARN, console
```

⑥ 启动 Spark 集群。执行 cd/usr/local/spark/sbin/命令，进入 Spark 安装包的/sbin 目录下，执行./start-all.sh 命令启动集群。

⑦ 启动 spark-sql。执行 cd/usr/local/spark/bin 命令进入 Spark 安装包的/bin 目录下，执行./spark-sql 命令来开启 Spark SQL CLI。在 Spark SQL CLI 中，我们可以直接执行 HiveQL 语句。

8.1.3　Spark SQL 与 Shell 交互

Spark SQL 已经集成在 Spark Shell 中，因此只要启动 Spark Shell，就可以使用 Spark SQL 的 Shell 交互接口。Spark SQL 对数据的查询分成了两个分支（接口）：SQLContext 和 HiveContext。HiveContext 继承了 SQLContext，因此，HiveContext 除了拥有 SQLContext 的特性，还拥有自身的特性。目前，Spark 从 Spark 1.1 开始，提供了两种语法解析器——SQL 语法解析器和 HiveQL 语法解析器。SQLContext 目前只支持 SQL 语法解析器，而 HiveContext 不仅支持 HiveQL 语法解析器，同时也支持 SQL 语法解析器。

Spark Shell 启动的过程中会初始化 SQLContext 对象为 SQLContext，此时，初始化的 SQLContext 对象既支持 SQL 语法解析器，也支持 HiveQL 语法解析器。也就是说，使用 SQLContext 可以执行 SQL 语句和 HQL 语句。

当然，读者也可以自己声明 SQLContext 对象，创建 SQLContext 对象的方式如下。通过这种方式创建的 SQLContext 只能执行 SQL 语句，不能执行 HQL 语句。

```
from pyspark import SparkContext,SparkConf

sqlContext = conf = SparkConf().setAppName("the apache sparksql")
```

8.2　DataFrame 基础操作

【任务描述】掌握 DataFrame 对象的创建方法和操作方法。

8.2.1 创建 DataFrame 对象

在 Spark SQL 中，开发人员可以非常便捷地将各种数据源数据转换为 DataFrame。

Python 示例代码如下，这段代码充分展现了 Spark SQL 中 DataFrame 数据源丰富多样和简单易用的特性。

```
# 从 Hive 中的 users 表构造 DataFrame
users = sqlContext.table("users")

# 加载 S3 上的 JSON 文件构造 DataFrame
logs = sqlContext.read.format("json").load("s3n:// path/to/data.json")

# 加载 HDFS 上的 Parquet 文件构造 DataFrame
clicks = sqlContext.read.format("parquet").load("hdfs:// path/to/data.parquet")

# 通过 JDBC 访问 MySQL 构造 DataFrame
comments = sqlContext.read.format("jdbc").option("url",
            "jdbc:mysql:// localhost/comments").option("user", "user").load()

# 将普通 RDD 转换为 DataFrame
rdd = sparkContext.textFile("article.txt") \
    .flatMap(lambda line: line.split()) \
    .map(lambda word: (word, 1)) \
    .reduceByKey(lambda a, b: a + b)
wordCounts = rdd.toDF(["word", "count"])

# 使用 RDD 的 toDF 函数创建 DataFrame
rdd = sc.parallelize([('Michael', 29), ('Andy', 30), ('Justin', 19)])
info = rdd.toDF()

# 将本地数据容器转换为 DataFrame
data = [("Alice", 21), ("Bob", 24)]
people = sqlContext.createDataFrame(data, ["name", "age"])

# 将 Pandas DataFrame 转换为 Spark DataFrame（Python API 特有功能）
sparkDF = sqlContext.createDataFrame(pandasDF)
```

Java 示例代码如下。

```
import org.apache.spark.api.java.JavaRDD;
import org.apache.spark.api.java.JavaSparkContext;
import org.apache.spark.sql.Dataset;
import org.apache.spark.sql.Row;
import org.apache.spark.sql.SparkSession;
import org.apache.spark.sql.jdbc.JdbcOptions;
import org.apache.spark.sql.sources.DataSourceRegister;
import org.apache.spark.sql.types.DataTypes;
import java.util.Arrays;
import java.util.List;
```

```
public class DataFrameExample {
    public static void main(String[] args) {
        SparkSession spark = SparkSession.builder()
                .appName("DataFrameExample")
                .getOrCreate();

        // 从 Hive 中的 users 表构造 DataFrame
        Dataset<Row> users = spark.table("users");

        // 加载 S3 上的 JSON 文件，构造 DataFrame
        Dataset<Row> logs = spark.read().format("json").load(
                            "s3n:// path/to/data.json");

        // 加载 HDFS 上的 Parquet 文件，构造 DataFrame
        Dataset<Row> clicks = spark.read().format("parquet").load(
                            "hdfs:// path/to/data.parquet");

        // 通过 JDBC 访问 MySQL 来构造 DataFrame
        Dataset<Row> comments = spark.read().format("jdbc")
                .option(JdbcOptions.JDBC_URL(), "jdbc:mysql:// localhost/comments")
                .option(JdbcOptions.JDBC_TABLE_NAME(), "user")
                .load();

        // 将普通 RDD 转换为 DataFrame
        List<String> lines = Arrays.asList("Hello Spark", "Hello World");
        JavaSparkContext jsc = new JavaSparkContext(spark.sparkContext());
        JavaRDD<String> rdd = jsc.parallelize(lines);
        Dataset<Row> wordCounts = spark.createDataFrame(rdd.map(line ->
                                line.split(" ")), DataTypes.createStructType
                                (Arrays.asList(DataTypes.createStructField
                                ("word", DataTypes.StringType, false))));

        // 使用 Java API，创建 DataFrame
        List<Row> rowData = Arrays.asList(RowFactory.create("Michael", 29),
                                    RowFactory.create("Andy", 30),
                                    RowFactory.create("Justin", 19));
        StructType schema = new StructType(new StructField[]{
                new StructField("name", DataTypes.StringType, false,
                                Metadata.empty()),
                new StructField("age", DataTypes.IntegerType, false,
                                Metadata.empty())
        });
        Dataset<Row> info = spark.createDataFrame(rowData, schema);
        // 将本地数据容器转换为 DataFrame
        List<Row> data = Arrays.asList(RowFactory.create("Alice", 21),
                        RowFactory.create("Bob", 24));
        StructType schema = new StructType(new StructField[]{
                new StructField("name", DataTypes.StringType, false,
                                Metadata.empty()),
```

```
                new StructField("age", DataTypes.IntegerType, false,
                                Metadata.empty())
        });
        Dataset<Row> people = spark.createDataFrame(data, schema);

        // 将 Pandas DataFrame 转换为 Spark DataFrame（需要 PySpark）
        // (在 Java 中无原生的等效实现，该功能是 Python API 特有的)
    }
}
```

由上述代码可见，从 Hive 表到外部数据源，API 所支持的各种数据源（JSON、Parquet、JDBC）、RDD 乃至各种本地数据集，都可以被方便快捷地进行加载，转换为 DataFrame。

从 Spark 2.0 版本开始，Spark 使用全新的 SparkSession 接口替代 Spark 1.6 中的 SQLContext 及 HiveContext 接口，实现数据的加载、转换、处理等功能。SparkSession 实现 SQLContext 及 HiveContext 的所有功能。

SparkSession 支持从不同的数据源中加载数据，把数据转换成 DataFrame，而且支持把 DataFrame 转换成 SQLContext 自身中的表，并使用 SQL 语句来操作数据。SparkSession 也提供了 HiveQL 及其他依赖于 Hive 的功能。

Python 示例代码如下。

```
# 创建 Sparksession 对象

from pyspark.sql import SparkSession

spark = SparkSession.builder.getorCreate()

# 通过 JSON 创建 DataFrame

df = spark.read.json ("file:///spark/examples/src/main/resources/people.json")

df.show()

# 通过 csv 创建 DataFrame

df = spark.read.csv ("file:///path/to/csv/file.csv")

df.show()

df.printSchema()

# 本地数据容器转变为 DataFrame

spark.createDataFrame([('Alice',1)]).collect()
```

Java 示例代码如下。

```
import org.apache.spark.sql.Dataset;
import org.apache.spark.sql.Row;
import org.apache.spark.sql.SparkSession;
```

```
public class DataFrameExample {
    public static void main(String[] args) {
        SparkSession spark = SparkSession.builder()
                .appName("DataFrameExample")
                .getOrCreate();

        // 通过 JSON 创建 DataFrame
        Dataset<Row> df = spark.read().json(
                        " file:///spark/examples/src/main/resources/people.json");
        df.show();

        // 通过 CSV 创建 DataFrame
        Dataset<Row> df = spark.read().csv("file:///path/to/csv/file.csv");
        df.show();
        df.printSchema();

        // 本地数据容器转变为 DataFrame
        List<Row> data = Arrays.asList(RowFactory.create("Alice", 1));
        StructType schema = new StructType(new StructField[]{

                new StructField("name", DataTypes.StringType,
                                false, Metadata.empty()),
                new StructField("age", DataTypes.IntegerType,
                                false, Metadata.empty())
        });
        Dataset<Row> df = spark.createDataFrame(data, schema);
        df.collect();
    }
}
```

下面具体讲解 crcateDataFrame(data, schema = None, samplingRatio = None, verifySchema = True)函数。createDataFrame()方法的参数及其含义如表 8-1 所示。

表 8-1　createDataFrame()方法的参数及其含义

参数	解释
data	指定数据
scherma	指定表结构
samplingRatio	指定用于类型推断的样本的比例。如果某列数据的类型不确定，则可以抽样 samplingRatio 的数据来判断该列数据是什么类型
verifySchema	验证每行的数据类型是否符合 schema 的定义

下面通过多个具体的示例来展示各类型的参数的用法，以加深读者的理解。

Python 示例代码如下。

```
from pyspark.sql import SparkSession
from pyspark.sql.types import *
```

```
# 创建 SparkSession 对象
spark = SparkSession.builder.getOrCreate()

# data 每行数据元组，没有指定 schema
spark.createDataFrame([('Alice', 1)]).show()

# data 每行数据是 dict，没有指定 schema
spark.createDataFrame([{'name': 'Alice', 'age': 1}]).show()

# data 用 rdd，没有指定 schema
rdd = spark.sparkContext.parallelize([('Alice', 1)])
spark.createDataFrame(rdd).show()

# data 每行数据用 Row，没有指定 schema
spark.createDataFrame([('Alice', 1)]).show()

# data 每行数据用元组，指定 schema 为列表
spark.createDataFrame([('Alice', 1)], ['name', 'age']).show()

# data 用 rdd，指定 schema 为列表
rdd = spark.sparkContext.parallelize([('Alice', 1)])
spark.createDataFrame(rdd, ['name', 'age']).show()

# data 用 rdd，指定 schema 为 StructField
schema = StructType([StructField("name", StringType(), True), StructField(
        "age", IntegerType(), True)])
rdd = spark.sparkContext.parallelize([('Alice', 1)])
df3 = spark.createDataFrame(rdd, schema)
df3.show()

# data 用 rdd，指定 schema 为"fieldName: fieldType"
rdd = spark.sparkContext.parallelize([{'name': 'Alice', 'age': 1}])
spark.createDataFrame(rdd, "name:string, age: int").show()

# data 用 rdd，只有一列,指定 schema 为"fieldType"
rdd = spark.sparkContext.parallelize([18])
spark.createDataFrame(rdd, "int").show()
```

Java 示例代码如下。

```
import org.apache.spark.sql.Dataset;
import org.apache.spark.sql.Row;
import org.apache.spark.sql.RowFactory;
import org.apache.spark.sql.SparkSession;
import org.apache.spark.sql.types.*;
import org.apache.spark.api.java.JavaRDD;
import org.apache.spark.api.java.JavaSparkContext;
import scala.Tuple2;
import java.util.Arrays;
import java.util.Collections;
```

```
public class DataFrameExample {

    public static void main(String[] args) {
        SparkSession spark = SparkSession.builder().master("local").appName("
                            SparkJavaExample").getOrCreate();
        JavaSparkContext jsc = new JavaSparkContext(spark.sparkContext());

        // data 每行数据元组，没有指定 schema
        Dataset<Row> df1 = spark.createDataFrame(Arrays.asList(new Tuple2<>
                            ("Alice", 1)), Tuple2.class);
        df1.show();

        // data 每行数据是 dict，没有指定 schema
        Dataset<Row> df2 = spark.createDataFrame(Arrays.asList(
                            Collections.singletonMap("name", "Alice")), Row.class);
        df2.show();

        // data 用 rdd，没有指定 schema
        JavaRDD<Tuple2<String, Integer>> rdd1 = jsc.parallelize
        (Collections.singletonList(new Tuple2<>("Alice", 1)));
        Dataset<Row> df3 = spark.createDataFrame(rdd1, Tuple2.class);
        df3.show();

        // data 每行数据用元组，指定 schema 为列表
        StructType schema1 = new StructType(new StructField[]{
            new StructField("name", DataTypes.StringType, true, Metadata.empty()),
            new StructField("age", DataTypes.IntegerType, true, Metadata.empty())
        });
        JavaRDD<Row> rddWithSchema1 = jsc.parallelize(
        Arrays.asList(RowFactory.create("Alice", 1)));
        Dataset<Row> df4 = spark.createDataFrame(rddWithSchema1, schema1);
        df4.show();


        // data 用 rdd，指定 schema 为 StructField
        StructType schema2 = new StructType(new StructField[]{
                new StructField("name", DataTypes.StringType, true,
                                Metadata.empty()),
                new StructField("age", DataTypes.IntegerType, true,
                                Metadata.empty())
        });
        JavaRDD<Row> rdd2 = jsc.parallelize(Collections.singletonList(
                            RowFactory.create("Alice", 1)));
        Dataset<Row> df5 = spark.createDataFrame(rdd2, schema2);
        df5.show();

        // data 用 rdd，指定 schema 为"fieldName: fieldType"
        JavaRDD<Row> rdd3 = jsc.parallelize(Collections.singletonList(
```

```
                        RowFactory.create("Alice", 1)));

        StructType schema3 = new StructType(new StructField[]{
                new StructField("name", DataTypes.StringType, true,
                                Metadata.empty()),
                new StructField("age", DataTypes.IntegerType, true,
                                Metadata.empty())
        });

        Dataset<Row> df6 = spark.createDataFrame(rdd3, schema3);
        df6.show();


        // data用rdd，只有一列，指定schema为"fieldType"
        JavaRDD<Integer> rdd4 = jsc.parallelize(Collections.singletonList(18));
        Dataset<Row> df7 = spark.createDataFrame(rdd4, Integer.class);
        df7.show();

        spark.stop();
    }
}
```

在大部分情况下，DataFrame被创建时需要定义schema、每一个字段名与数据类型，这样后续操作可以用字段名进行统计。默认DataFrame API已经定义很多类似于SQL的方法，例如select().groupby().count()等，我们可以使用这些方法进行统计。

8.2.2 DataFrame查看数据

DataFrame派生于RDD类，因此和RDD类似，DataFrame只有在触发行动操作时才会进行计算。DataFrame提供了很多常用的查看及获取数据的操作函数（方法），如表8-2所示。

表8-2 DataFrame常用的操作函数（方法）

函数（方法）	描述
printSchema()	打印数据模式
show()	查看数据
first()、head()、take()	获取若干行数据
collect()	获取所有数据

下面为展示DataFrame查看数据的操作，我们加载一份数据。

```
from pyspark import SparkContext
from pyspark.sql import SparkSession
from pyspark.sql import Row
```

```
sc = SparkContext("local","app")
sqlContext = SparkSession.builder.getOrCreate()
RemSDR = sc.textFile("./sales_data.csv")
print(RemSDR.count()) # 计数

# 去表头
salesRDD = RemSDR.map(lambda x:x.split(","))
title = salesRDD.take(1)[0]
salesTRDD = salesRDD.filter(lambda line:line != title)
print(salesTRDD.count()) # 计数

sale_Rows = salesTRDD.map(lambda p:
             Row(
              ORDERNUMBER = p[0],
              SALES = p[1],
              ORDERDATE = p[2],
              YEAR_ID = p[3],
              PRODUCTCODE = p[4],
              CITY = p[5],
              POSTALCODE = p[6],
              STATE = p[7]
           )
)

# 创建DataFrame
sale_df = sqlContext.createDataFrame(sale_Rows)
```

1．printSchema：打印数据模式

在创建完 DataFrame 之后，一般都会查看 DataFrame 中的数据模式。我们可以通过 printSchema()函数来查看数据模式，该函数会打印出列的名称和类型。示例如下。

```
# 显示Schema
sale_df.printSchema()


# 结果显示
root
 |-- ORDERNUMBER: string (nullable = true)
 |-- SALES: string (nullable = true)
 |-- ORDERDATE: string (nullable = true)
 |-- YEAR_ID: string (nullable = true)
 |-- PRODUCTCODE: string (nullable = true)
 |-- CITY: string (nullable = true)
 |-- POSTALCODE: string (nullable = true)
 |-- STATE: string (nullable = true)
```

2．show：查看数据

打印完数据模式之后，我们还需要查看加载到 DataFrame 中的数据是否正确。从创建的 DataFrame 中对数据进行采样的方法有很多种，其中最简单的是 show()方法，如表 8-3 所示。

表 8-3　show 方法

方法	解释
show()	显示前 20 条记录
show(numRows: Int)	显示 numRows 条记录
show(truncate: Boolean)	是否最多只显示 20 个字符，默认为 True
show(numRows:Int, truncate: Boolean)	显示 numRows 条记录，并设置过长字符串的显示格式

下面代码使用 show()方法查看 DataFrame 对象 sale_df 中的数据。show()方法与 show(truncate = True)方法一样，只显示前 20 条记录，并且最多只显示 20 个字符。若要显示所有字符，则使用 show(truncate = False)方法。

```
sale_df.show()

sale_df.show(truncate = False)
```

若想查看前 n（n<20）行记录，则可以使用 show(numRows:Int)方法，例如查看 sale_df 的前 10 行记录，代码如下。

```
sale_df.show(10)
```

3．first/head/take：获取若干行记录

要获取 DataFrame 若干行记录，除了使用 show()方法外，还可以使用 first()、head()、take()等方法。获取 DataFrame 若干行记录的方法及其解释如表 8-4 所示。

表 8-4　获取 DataFrame 若干行记录的方法及其解释

方法	解释
first()	获取第 1 行记录
head(n:Int)	获取前 n 行记录
take(n:Int)	获取前 n 行记录

first()和 head()方法的功能相同，以 Row 或者 Array[Row]的形式返回一行或多行数据。take()方法会将获得的数据返回到驱动器端。这 3 个方法的使用示例如下。

```
sale_df.show()

sale_df.head(3)

sale_df.take(3)
```

4．collect：获取所有数据

不同于 show()方法，collect()方法可以获取 DataFrame 中的所有数据，并返回一个 Array 对象。下列代码是 collect()方法的用法。

```
sale_df.collect()
```

8.2.3 DataFrame 查询操作

DataFrame 查询有两种方法，第一种方法是将 DataFrame 注册成临时表，然后通过 SQL 语句进行查询。具体操作如下。

```
#创建 sale_table 临时表

sale_df.registerTempTable("sale_table")
```

第二种方法是直接在 DataFrame 对象上进行查询。DataFrame 的查询操作也是一个“懒操作”，仅仅生成一个查询计划，只有触发行动操作，才会进行计算并返回查询结果。

1. 选择字段

以下示例是在 Java 和 Python 中使用 DataFrame 和 Spark SQL 来选择 SALES 和 ORDERDATE 字段的方法。

（1）使用 DataFrame 选择字段

在 Spark 的 DataFrame API 中，select()方法的作用是从 DataFrame 中选取一个或多个列，并返回一个新的 DataFrame。通过使用 select()方法，开发人员可以轻松地从大型 DataFrame 中提取他们所需的特定列。

Python 示例代码如下。

```
from pyspark.sql import SparkSession

spark = SparkSession.builder \
    .appName("FieldSelectionPython") \
    .master("local") \
    .getOrCreate()

df = spark.read.json("path_to_your_data.json")

# 选择 SALES 和 ORDERDATE 字段，并显示
selected_df = df.select("SALES", "ORDERDATE")
selected_df.show()
```

Java 示例代码如下。

```
import org.apache.spark.sql.Dataset;
import org.apache.spark.sql.Row;
import org.apache.spark.sql.SparkSession;

public class FieldSelectionJava {
    public static void main(String[] args) {
        SparkSession spark = SparkSession.builder()
                .appName("FieldSelectionJava")
                .master("local")
                .getOrCreate();

        Dataset<Row> df = spark.read().json("path_to_your_data.json");
```

```
        // 选择SALES和ORDERDATE字段，并显示
        Dataset<Row> selectedDf = df.select("SALES", "ORDERDATE");
        selectedDf.show();
    }
}
```

（2）使用 Spark SQL 选择字段

Python 示例代码如下。

```
from pyspark.sql import SparkSession

spark = SparkSession.builder \
    .appName("SparkSQLSelectionPython") \
    .master("local") \
    .getOrCreate()

df = spark.read.json("path_to_your_data.json")

# 使用Spark SQL选择SALES和ORDERDATE字段
df.createOrReplaceTempView("sales")

result = spark.sql("SELECT SALES, ORDERDATE FROM sales")
result.show()
```

Java 示例代码如下。

```
import org.apache.spark.sql.Dataset;
import org.apache.spark.sql.Row;
import org.apache.spark.sql.SparkSession;

public class SparkSQLSelectionJava {
    public static void main(String[] args) {
        SparkSession spark = SparkSession.builder()
                .appName("SparkSQLSelectionJava")
                .master("local")
                .getOrCreate();

        Dataset<Row> df = spark.read().json("path_to_your_data.json");

        // 使用Spark SQL选择SALES和ORDERDATE字段
        df.createOrReplaceTempView("sales");

        Dataset<Row> result = spark.sql("SELECT SALES, ORDERDATE FROM sales");
        result.show();
    }
}
```

2. 增加字段

（1）使用 DataFrame 增加计算字段

Python 示例代码如下。

```
from pyspark.sql.functions import col
```

```
# Assuming sales_data has a column "sale_year"
calculated_df = sales_data.withColumn("warranty_years",2023 - col("sale_year"))
```

Java 示例代码如下。

```
import org.apache.spark.sql.Dataset;
import org.apache.spark.sql.Row;
import org.apache.spark.sql.functions;

Dataset<Row> calculatedDF = salesData.withColumn("warranty_years",functions.expr("
2023 - sale_year"));
```

（2）使用 Spark SQL 增加计算字段

Python 示例代码如下。

```
sales_data.createOrReplaceTempView("sales")
spark.sql("SELECT *, 2023 - sale_year AS warranty_years FROM sales")
```

Java 示例代码如下。

```
salesData.createOrReplaceTempView("sales");
Dataset<Row> result = spark.sql("SELECT *, 2023 - sale_year AS
warranty_years FROM sales");
```

3．条件查询

假设筛选 2018 年之后的销售数据。

（1）使用 DataFrame 筛选数据

Python 示例代码如下。

```
filtered_df = sales_data.filter(col("sale_year") > 2018)
```

Java 示例代码如下。

```
Dataset<Row> filteredDF = salesData.filter(
                        functions.col("sale_year").gt(2018));
```

（2）使用 Spark SQL 筛选数据

Python 示例代码如下。

```
spark.sql("SELECT * FROM sales WHERE sale_year > 2018")
```

Java 示例代码如下。

```
Dataset<Row> result = spark.sql("SELECT * FROM sales WHERE sale_year > 2018");
```

4．数据排序

假设按照销售年份对数据进行排序。

（1）使用 DataFrame 进行数据排序

Python 示例代码如下。

```
sorted_df = sales_data.orderBy("sale_year")
```

Java 示例代码如下。

```
Dataset<Row> sortedDF = salesData.orderBy("sale_year");
```

（2）使用 Spark SQL 进行数据排序

Python 示例代码如下。

```
spark.sql("SELECT * FROM sales ORDER BY sale_year")
```

Java 示例代码如下。

```
Dataset<Row> result = spark.sql("SELECT * FROM sales ORDER BY sale_year");
```

5. 数据去重

假设我们想删除所有重复的销售记录。

（1）使用 DataFrame 进行数据去重

Python 示例代码如下。

```
distinct_df = sales_data.distinct()
```

Java 示例代码如下。

```
Dataset<Row> distinctDF = salesData.distinct();
```

（2）使用 Spark SQL 进行数据去重

Python 示例代码如下。

```
spark.sql("SELECT DISTINCT * FROM sales")
```

Java 示例代码如下。

```
Dataset<Row> result = spark.sql("SELECT DISTINCT * FROM sales");
```

6. 数据分组统计

假设我们想按照销售年份对数据进行分组，并计算每年的销售数量。

（1）使用 DataFrame 进行数据分组并计算年销量

Python 示例代码如下。

```
from pyspark.sql.functions import count

grouped_df = sales_data.groupBy("sale_year").agg(count("*").alias(
            "count_per_year"))
```

Java 示例代码如下。

```
import static org.apache.spark.sql.functions.*;

Dataset<Row> groupedDF = salesData.groupBy("sale_year").agg(count("*").as(
                        "count_per_year"));
```

（2）使用 Spark SQL 进行数据分组并计算年销量

Python 示例代码如下。

```
spark.sql("SELECT sale_year, COUNT(*) as count_per_year FROM sales GROUP BY
sale_year")
```

Java 示例代码如下。

```
Dataset<Row> result = spark.sql("SELECT sale_year, COUNT(*) as
                     count_per_year FROM sales GROUP BY sale_year");
```

7. 数据连接

sale_table 有 POSTALCODE 字段为空的销售记录，因此需连接 Zipssortedbycitystate 表，进而完善销售记录。

（1）加载数据并注册成表

Python 示例代码如下。

```
from pyspark.sql import SparkSession

spark = SparkSession.builder \
    .appName("DataLoadingPython") \
    .master("local") \
```

```
    .getOrCreate()

sale_table = spark.read.json("path_to_sale_table_data.json")
zip_table = spark.read.json("path_to_Zipssortedbycitystate_data.json")

sale_table.createOrReplaceTempView("sale_table")
zip_table.createOrReplaceTempView("zip_table")
```

Java 示例代码如下。

```
import org.apache.spark.sql.Dataset;

import org.apache.spark.sql.Row;
import org.apache.spark.sql.SparkSession;

public class DataLoadingJava {
    public static void main(String[] args) {
        SparkSession spark = SparkSession.builder()
                .appName("DataLoadingJava")
                .master("local")
                .getOrCreate();

        Dataset<Row> sale_table = spark.read().json(
                                  "path_to_sale_table_data.json");
        Dataset<Row> zip_table = spark.read().json(
                                  "path_to_Zipssortedbycitystate_data.json");

        sale_table.createOrReplaceTempView("sale_table");
        zip_table.createOrReplaceTempView("zip_table");
```

（2）使用 DataFrame 进行数据连接

Python 示例代码如下。

```
joined_df = sale_table.join(zip_table, sale_table["POSTALCODE"] ==
            zip_table["ZIP Code"], "left_outer")
joined_df.show()
```

Java 示例代码如下。

```
Dataset<Row> joinedDf = sale_table.join(zip_table, sale_table.col(
                        "POSTALCODE").equalTo(zip_table.col("ZIP Code")),
                        "left_outer");
joinedDf.show();
```

（3）使用 Spark SQL 进行数据连接

Python 示例代码如下。

```
result = spark.sql("SELECT * FROM sale_table LEFT OUTER JOIN Zipssortedbycitystate
ON sale_table.POSTALCODE = Zipssortedbycitystate.`ZIP Code`")
result.show()
```

Java 示例代码如下。

```
Dataset<Row> result = spark.sql("SELECT * FROM sale_table LEFT OUTER JOIN
Zipssortedbycitystate ON sale_table.POSTALCODE = Zipssortedbycitystate.`ZIP Code`");
result.show();
```

8.2.4 DataFrame 输出操作

新建实例如下。

在/usr/local/spark/examples/src/main/resources/目录下，有一个样例数据 people.json 文件。people.json 文件的内容如下。

```
{"name":"Michael"}

{"name":"Andy", "age":30}

{"name":"Justin", "age":19}
```

建立一个 DataFrame，命令如下。

```
>>> df = spark.read.json(
      "file:///usr/local/spark/examples/src/main/resources/people.json")

>>> df.show()

+----+-------+
| age|   name|
+----+-------+
|null|Michael|
|  30|   Andy|
|  19| Justin|
+----+-------+
```

我们可以使用 spark.write 操作，把一个 DataFrame 保存成不同格式的文件。例如，把一个名称为 df 的 DataFrame 保存到不同格式文件中的方法如下。

```
df.write.text("people.txt")

df.write.json("people.json")

df.write.parquet("people.parquet")
```

我们也可以使用以下格式的语句。

```
df.write.format("text").save("people.txt")

df.write.format("json").save("people.json")

df.write.format ("parquet").save("people.parquet")
```

下面以 people.json 文件为基础，创建一个 DataFrame，其名称为 peopleDF。我们把 peopleDF 保存到另一个 JSON 文件中，并从 peopleDF 中选取一个列（name 列），把该列数据保存到一个文本文件中。

```
>>> peopleDF = spark.read.format("json").\

... load("file:///usr/local/spark/examples/src/main/resources/people.json")

>>> peopleDF.select("name", "age").write.format("json").\
```

```
... save("file:///usr/local/spark/mycode/sparksql/newpeople.json")

>>> peopleDF.select("name").write.format("text").\

... save("file:///usr/local/spark/mycode/sparksql/newpeople.txt")
```

这时生成了一个名称为 newpeople.json 的目录（不是文件）和一个名称为 newpeople.txt 的目录（不是文件）。

8.3 Spark SQL 与 MySQL 的交互

【任务描述】掌握 Spark SQL 与 MySQL 的交互方法，能够使用 Spark SQL 读/写数据库。

1．数据准备

使用 Spark SQL 读/写数据库，先要在 Linux 中启动 MySQL 数据库，命令如下。

```
$service mysql start
$mysql -u root -p
# 屏幕会提示你输入密码
```

输入以下 SQL 语句，完成数据库和表的创建。

```
mysql> create database spark;
mysql> use spark;
mysql> create table student (id int(4),name char(20),gender char(4),age int(4));
mysql> insert into student values(1,'Xueqian','F',23);
mysql> insert into student values(2,'Weiliang','M',24);
mysql> select * from student;
```

2．编写代码

Python 示例代码如下。

下载 MySQL 的 JDBC 驱动（例如 mysql-connector-java-5.1.40.tar.gz），把该驱动存储到 spark 的安装目录/usr/local/spark/jars 下。

启动 PySpark，命令如下。

```
$cd /usr/local/spark
$./bin/pyspark
```

执行以下命令连接数据库，读取并显示数据。

```
>>>jdbcDF = spark.read \
    .format("jdbc") \
    .option("driver","com.mysql.jdbc.Driver") \
    .option("url", "jdbc:mysql://localhost:3306/spark") \
    .option("dbtable", "student") \
    .option("user", "root") \
    .option("password", "123456") \
    .load()

>>> jdbcDF.show()
```

```
+---+--------+------+---+
| id| name|gender|age|
+---+--------+------+---+
| 1| Xueqian| F| 23|
| 2| Weiliang| M| 24|
+---+--------+------+---+
```

在 MySQL 数据库中创建一个名称为 spark 的数据库，并创建一个名称为 student 的数据表。

Java 示例代码如下。为了在 Java 中使用 Spark 的 DataFrame API 连接到 MySQL 数据库并读取数据，我们需要先确保已经正确安装 MySQL 的 JDBC 驱动并将其放在适当的位置。

```
import org.apache.spark.sql.Dataset;
import org.apache.spark.sql.Row;
import org.apache.spark.sql.SparkSession;
import java.util.Properties;

public class SparkMySQLJava {
    public static void main(String[] args) {
        SparkSession spark = SparkSession.builder()
                .appName("SparkMySQLJava")
                .master("local")
                .getOrCreate();

        // 定义连接属性
        String url = "jdbc:mysql://localhost:3306/spark";
        Properties connectionProperties = new Properties();
        connectionProperties.put("driver", "com.mysql.jdbc.Driver");
        connectionProperties.put("user", "root");
        connectionProperties.put("password", "123456");

        // 使用 DataFrame API 读取数据
        Dataset<Row> jdbcDF = spark.read()
                .jdbc(url, "student", connectionProperties);

        // 显示数据
        jdbcDF.show();
    }
}
```

注意：确保在运行 Java 程序之前，你的 CLASSPATH 或项目的依赖管理系统（例如 Maven 或 Gradle）包含了 MySQL 的 JDBC 驱动。

3. 查看数据

创建数据库后，查看数据库内容，命令如下。

```
Mysql> use spark;
Database changed

mysql>select*from student;
+------+---------+-------+------+
| id | name | gender|age |
+------+---------+-------+------+
```

```
|   1|Xueqian|F   | 23|
|   2|Weiliang|M   | 24|
+------+----------+--------+------+
2 rows in set(0.00 seec)
```

然后开始编写程序，向 spark.student 表中插入两条记录，具体如下。

```
# !/usr/bin/env python3
from pyspark.sql import Row
from pyspark.sql.types import *
from pyspark import SparkContext,SparkConf
from pyspark.sql import SparkSession

spark = SparkSession.builder.config(conf = SparkConf()).getOrCreate()

# 下面设置模式信息
schema = StructType([StructField("id", IntegerType(), True), \
StructField("name", StringType(), True), \
StructField("gender", StringType(), True), \
StructField("age", IntegerType(), True)])

# 下面设置两条数据，表示两个学生的信息
studentRDD = spark \
.sparkContext \
.parallelize(["3 Rongcheng M 26","4 Guanhua M 27"]) \
.map(lambda x:x.split(""))


# 下面创建 Row 对象，每个 Row 对象都是 rowRDD 中的一行
rowRDD = studentRDD.map(lambda p:Row(int(p[0].strip()), p[1].strip(),
         p[2].strip(), int(p[3].strip())))
# 建立 Row 对象和模式之间的对应关系，也就是把数据和模式对应起来
studentDF = spark.createDataFrame(rowRDD, schema)

# 写入数据库
prop = {}
prop['user'] = 'root'
prop['password'] = '123456'
prop['driver'] = "com.mysql.jdbc.Driver"
studentDF.write.jdbc("jdbc:mysql://localhost:3306/spark",'student','append',
prop)
```

查看 MySQL 数据库中的 spark.student 表发生了哪些变化，命令如下。

```
Mysql> select * from student;


+------+-----------+--------+------+
| id | name | gender | age |
+------+-----------+--------+------+
| 1 | Xueqian | F | 23 |
| 2 | Weiliang | M | 24 |
| 3 | Rongcheng | M | 26 |
| 4 | Guanhua | M | 27 |
```

```
+-------+-----------+--------+------+
4 rows in set (0.00 sec)
```

8.4 本章小结

本章详细介绍了 Spark SQL 和 DataFrame 的相关内容，通过对比突出了它们的独特性和差异。通过一个销售数据分析案例，我们深入了解了字段计算、条件查找、数据排序、数据去重、数据分组统计、数据连接等常见功能。在 Spark SQL 中，DataFrame 可以被看作关系数据库中的一张表，我们可以通过 DataFrame 的 API 实现对其的查询操作。此外，本章还展示了如何使用 Spark SQL 与 MySQL 进行数据交互。

第 9 章 Spark Streaming：实时计算框架

本章的研究对象是 Spark Streaming。它是 Spark 的一个流计算组件，其功能是实时处理流式数据，例如，通过分析购物网站实时产生的点击事件，为商品提供个性化的推荐，进而提高经济回报。

【学习目标】

1．了解 Spark Streaming 及其核心要点。

2．熟悉 DStream 模型的基础概念，以及转换、窗口、输出等操作。

9.1 Spark Streaming 概述

【任务描述】了解 Spark Streaming 的基本概念和运行原理，掌握静态数据与流数据的差异及其计算方法，熟悉 Spark Streaming 的设计思路。

9.1.1 Spark Streaming 应用场景

在 Spark Streaming 框架中，数据的处理单位采用的是批量方式，而非单个数据项的方式。尽管数据的采集是按单个数据项进行的，但 Spark Streaming 会设置一个批处理间隔，确保数据在达到特定数量后再进行统一处理。这个批处理间隔不仅是 Spark Streaming 的核心概念和主要参数，还直接决定了 Spark Streaming 的作业提交频率、数据处理时延，会对数据处理的吞吐量和整体性能产生影响。Spark Stream 处理流程如图 9-1 所示。

图 9-1 Spark Stream 处理流程

9.1.2 流计算概述

1．静态数据和流数据

静态数据是指在运行过程中主要用于控制或参考的数据，它们在很长的一段时间内基本保持不变。流数据是一组顺序、大量、快速、连续到达的数据序列。一般情况下，数据流是随时间持续增长的动态数据集合。流数据可应用于网络监控、传感器网络、航空航天、气象测控、金融服务等领域。流数据具有以下特征。

① 数据快速持续到达，潜在数据规模也许是无穷无尽的。

② 数据来源众多，格式复杂。

③ 数据量大，一旦经过处理，要么被丢弃，要么被归档存储。

④ 注重数据的整体价值，不过分关注个别数据。

⑤ 数据可能出现乱序或者不完整，系统无法控制将要处理的新到达的数据元素的顺序。

2．批量计算和实时计算

对于静态数据和流数据，存在两种完全不同的计算模式：批量计算和实时计算。批量计算和实时计算模型如图 9-2 所示。批量计算主要针对静态数据进行，这意味着在开始计算之前，所有数据都已经准备就绪，这种计算方式主要用于数据挖掘和验证业务模型。相对地，流数据并不适合采用批量计算方式，因为其特性使得它不适合用传统的关系模型来建模，因此，流数据的处理必须采用实时计算，其响应时间通常为秒级。当数据量较少时，利用实时计算并不会产生问题，但是，在大数据时代，复杂的数据格式、多样的数据来源及巨大的数据量为实时计算带来了巨大的挑战。为此，专门针对流数据的实时计算——流计算应运而生。流计算：实时获取来自不同数据源的海量数据，经过实时分析处理，获得有价值的信息。

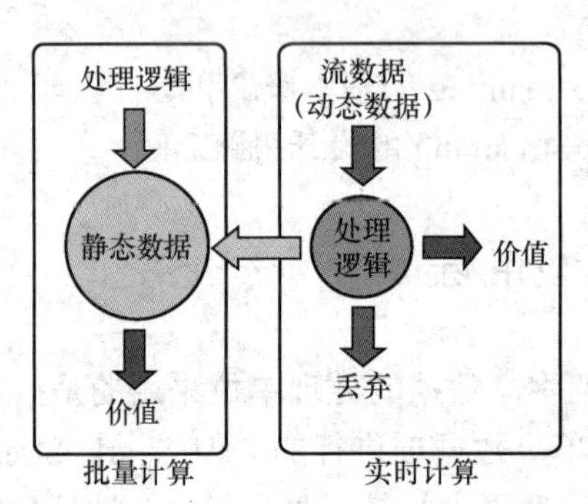

图 9-2　批量计算和实时计算模型

3．流计算的概念

流计算的核心理念是：数据的价值会随着时间的推移而逐渐减少，例如用户点击流的数据。因此，对于新出现的事件，我们应当立即处理，而不是将其存储以后再进行批量处理。为了能够实时处理这些流数据，我们需要的是一个具有低时延、可扩展和高可靠的处理引擎。一个高效的流计算系统应满足以下要求。

① 高性能：即使是大数据（例如每秒产生几十万条数据），也能够迅速处理。

② 海量式处理：能够处理 TB 甚至 PB 级别的数据。

③ 实时性：确保有很低的时延，能达到秒级甚至毫秒级。

④ 分布式架构：作为支持大数据的基础架构，应能够进行无缝扩展。

⑤ 易用性：支持快速的开发和部署。

⑥ 可靠性：确保流数据的可靠处理。

4．流计算框架

当前业界产生了许多专门的流数据实时计算系统来满足各自的需求。目前有 3 类常见的流计算框架和平台。

① 商业级：IBM InfoSphere Streams 和 IBM StreamBase。

② 较为常见的是开源流计算框架，它们中间的代表如下。

Twitter Storm：免费、开源的分布式实时计算系统，可简单、高效、可靠地处理大量的流数据。

Yahoo! S4（Simple Scalable Streaming System）：开源流计算平台，是通用的、分布式的、可扩展的、分区容错的、可插拔的流式系统。

③ 公司为支持自身业务开发的流计算框架，例如 Facebook Puma、DStream（百度）、银河流数据处理平台（淘宝）。

5．流计算处理流程

（1）概述

传统的数据处理流程需要先采集数据并将数据存储在关系数据库等数据管理系统中，之后由用户通过查询操作和数据管理系统进行交互。传统的数据处理流程如图 9-3 所示。

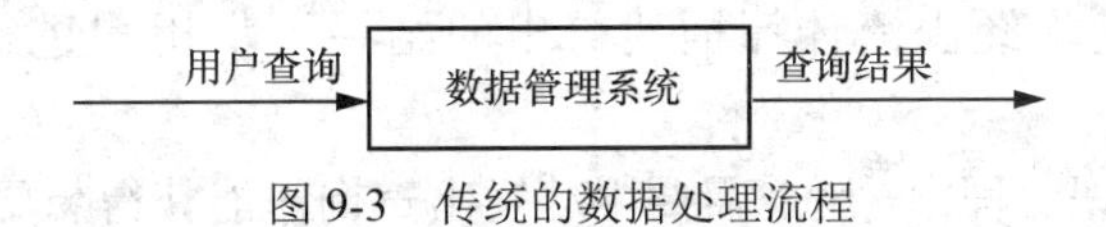

图 9-3　传统的数据处理流程

传统的数据处理流程隐含了两个前提：① 存储的数据是旧的，存储的静态数据是过去某一时刻的快照，这些数据在查询时可能已不具备时效性了；② 需要用户主动发出查询来获取结果。

流计算的处理流程一般包含 3 个阶段：数据实时采集、数据实时计算、实时查询服务。流计算处理流程如图 9-4 所示。

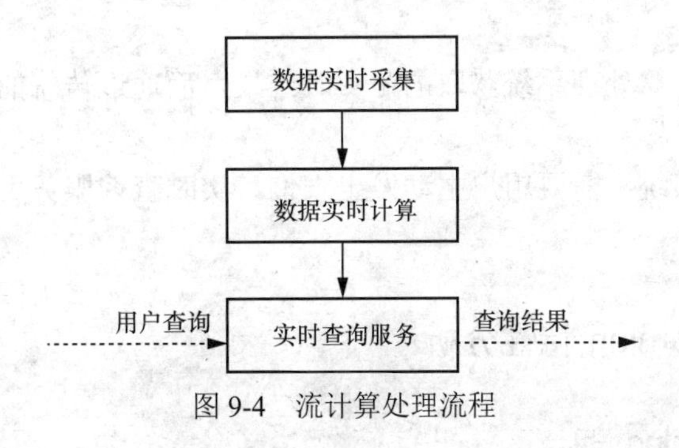

图 9-4　流计算处理流程

（2）数据实时采集

数据实时采集阶段通常采集多个数据源的海量数据，这需要数据处理系统能够具有实时性、低时延与稳定可靠的特点。

以日志数据为例，由于分布式集群的广泛应用，数据被分散存储在不同的机器上，因此系统需要实时汇总来自不同机器的日志数据。

目前，许多互联网公司发布的开源分布式日志采集系统均可满足每秒数百兆字节（MB）的数据采集和传输需求，例如 Facebook 的 Scribe、LinkedIn 的 Kafka、淘宝的 TimeTunnel、基于 Hadoop 的 Chukwa 和 Flume。

（3）数据实时计算

数据实时计算阶段对采集的数据进行实时的分析和计算，并实时反馈结果。

经流计算处理系统处理后的数据可视情况进行存储，以便之后再用于分析和计算。在对时效性要求较高的场景中，处理后的数据也可以直接丢弃。数据实时计算流程如图 9-5 所示。

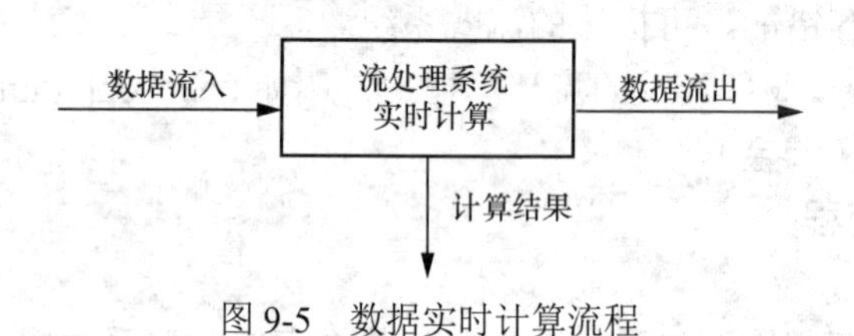

图 9-5　数据实时计算流程

（4）实时查询服务

实时查询服务是指经由流计算框架得出的结果可供用户进行实时查询、展示或存储。

在传统的数据处理流程中，用户需要主动发出查询才能获得想要的结果。而在流计算处理流程中，实时查询服务可以不断更新结果，并将用户所需的结果实时推送给用户。

虽然通过对传统的数据处理系统进行定时查询也可以实现实时更新结果和结果推送，但通过这样的方式所获取的结果仍然是根据过去某一时刻的数据得出来的，与实时结果有着本质上的区别。

综上所述，流计算处理系统与传统的数据处理系统有以下不同。

① 流计算处理系统处理的是实时的数据，而传统的数据处理系统处理的是预先存储好的静态数据。

② 用户通过流计算处理系统获取的是实时结果，而通过传统的数据处理系统获取的是过去某一时刻的结果。

③ 流计算处理系统不需要用户主动发出查询，实时查询服务可以主动将实时结果推送给用户。

9.1.3　Spark Streaming 特性分析

1．Spark Streaming 设计

Spark Streaming 可整合多种输入数据源，例如 Kafka、Flume、HDFS，甚至是普通的 TCP 套接字。经处理后的数据可存储至文件系统、数据库，或显示在仪表盘中。Spark Streaming 支持的输入、输出数据源如图 9-6 所示。

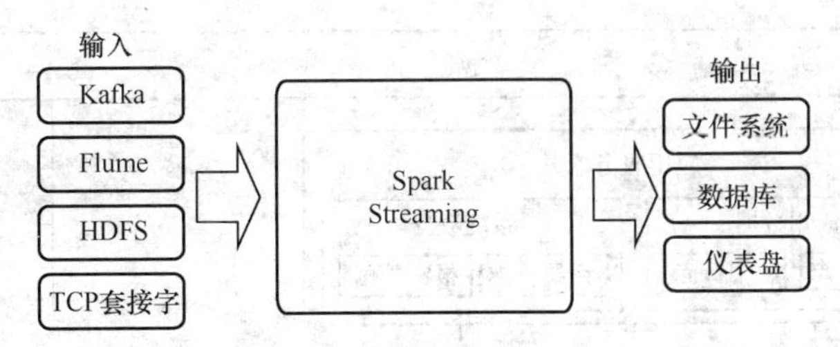

图 9-6 Spark Streaming 支持的输入、输出数据源

Spark Streaming 的基本原理是将实时输入的数据流以时间片（秒级）为单位进行拆分，由 Spark 引擎以类似批处理的方式处理每个时间片数据。Spark Streaming 执行流程如图 9-7 所示。

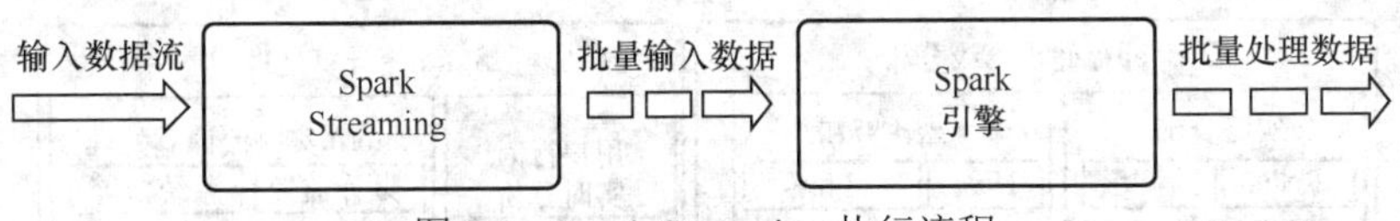

图 9-7 Spark Streaming 执行流程

Spark Streaming 最主要的抽象是离散化数据流（Discretized Stream，DStream），表示连续不断的数据流。在内部实现上，Spark Streaming 的输入数据按照时间片（例如以 1s 为单位）分成多段，每一段数据转换为 Spark 中的 RDD，这些分段就是 DStream，并且对 DStream 的操作最终转变为对相应的 RDD 操作。DStream 操作如图 9-8 所示。

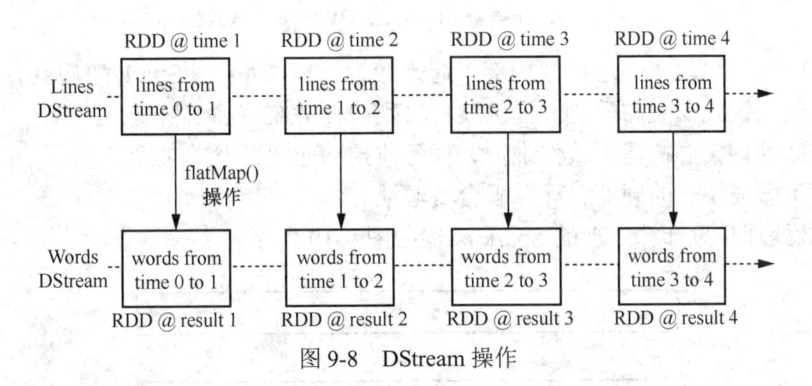

图 9-8 DStream 操作

2．Spark Streaming 与 Storm 的对比

Spark Streaming 和 Storm 最大的区别在于 Spark Streaming 无法实现毫秒级的流计算，而 Storm 可以实现毫秒级响应。

Spark Streaming 构建在 Spark 上有两方面因素：一方面是因为 Spark 的低时延执行引擎（100 + ms）可以用于实时计算；另一方面，相比于 Storm，RDD 数据集更易于做高效的容错处理。

Spark Streaming 采用的小批量处理的方式使它可以同时兼容批量和实时数据处理的逻辑和算法，因此，更适用于一些需要历史数据和实时数据联合分析的特定应用场合。

3．从"Hadoop + Storm"架构转向 Spark 架构

采用 Hadoop + Storm 架构的案例如图 9-9 所示。

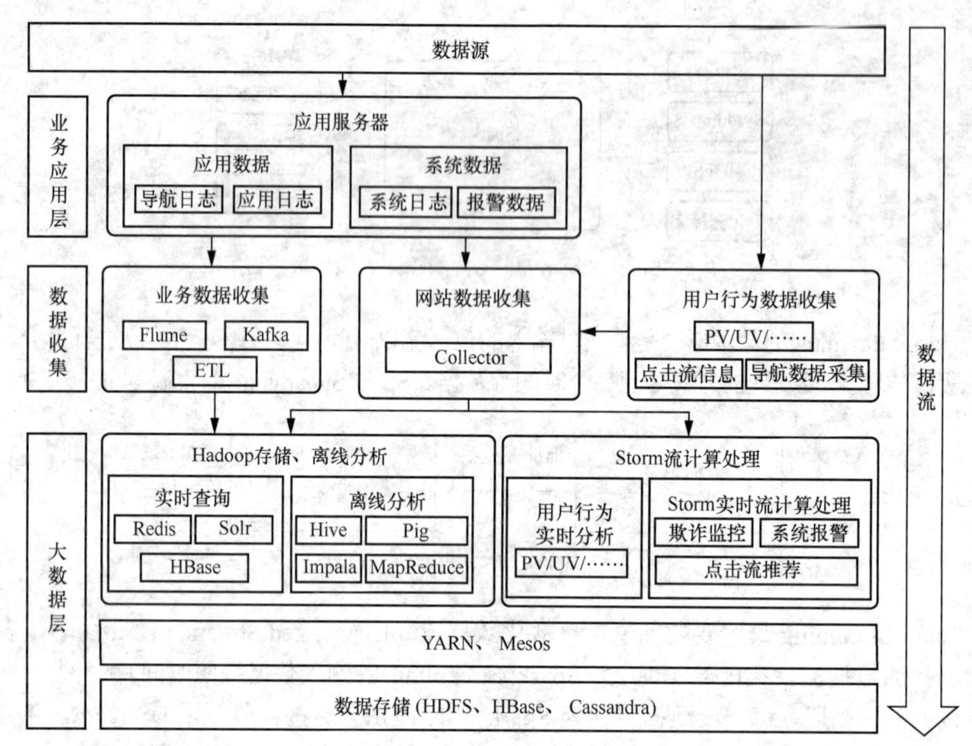

图 9-9　采用 Hadoop+Storm 架构的案例

注：PV——浏览次数；UV——浏览人数

从图 9-9 中可以看出，这种架构部署起来较为烦琐，而采用 Spark 架构具有以下优点。

① 实现一键式安装和配置，提供线程级别的任务监控和告警。

② 降低硬件集群构建、软件维护、任务监控和应用开发的难度。

③ 便于做成统一的硬件、计算平台资源池。

用于批处理和流计算处理的 Spark 架构如图 9-10 所示。

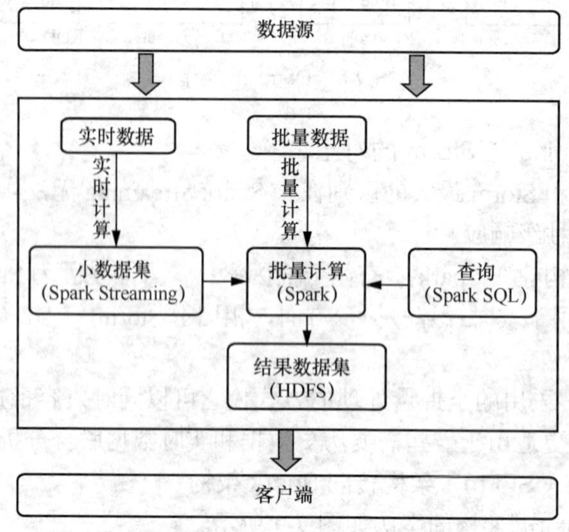

图 9-10　用于批处理和流计算处理的 Spark 架构

9.2 DStream 编程模型基础

【任务描述】了解 DStream 的基本概念，学会运行 Spark Streaming，掌握 DStream 的输入、转换、窗口、输出等操作。

9.2.1 DStream 概述

1. Spark Streaming 工作机制

Spark Streaming 工作机制如图 9-11 所示。

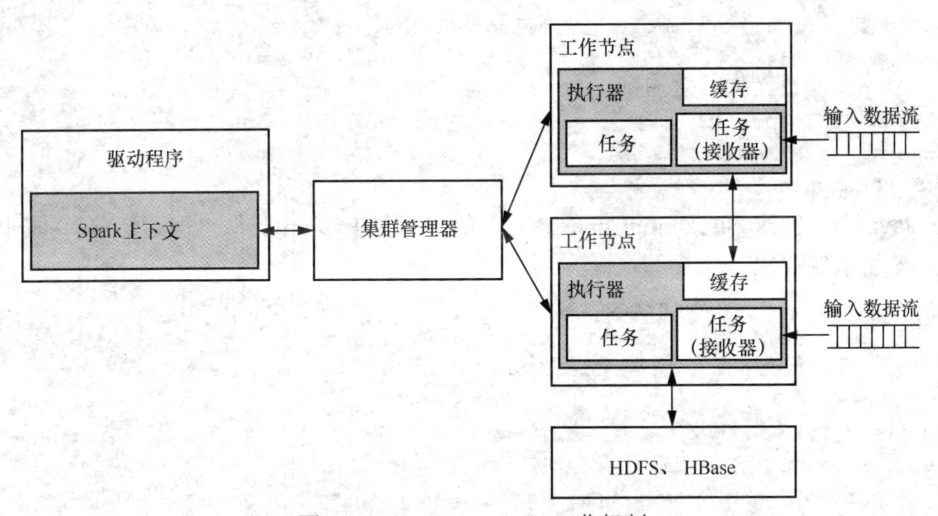

图 9-11　Spark Streaming 工作机制

在 Spark 流计算处理中，会有一个组件任务接收器，作为一个长期运行的任务运行在一个执行器上。

每个任务接收器都会负责一个输入数据流（例如从文件中读取数据的文件流，又如套接字流，或者从 Kafka 中读取一个输入流等）。

Spark 流计算处理通过输入数据流与外部数据源进行连接，读取相关数据。

2. 编写 Spark Streaming 程序的基本步骤

编写 Spark Streaming 程序的基本步骤如下。

步骤 1：通过创建输入 DStream 来定义输入源。

步骤 2：通过对 DStream 应用转换操作和输出操作来定义流计算。

步骤 3：用 streamingContext.start()方法来开始接收数据和处理流程。

步骤 4：通过 streamingContext.awaitTermination()方法来等待处理结束（手动结束或因为错误而结束）。

步骤 5：可以通过 streamingContext.stop()方法来手动结束流计算进程。

3．创建 StreamingContext 对象

如果要运行一个 Spark Streaming 程序，就需要首先生成一个 StreamingContext 对象，该对象是 Spark Streaming 程序的主入口。

可以从一个 SparkConf 对象中创建一个 StreamingContext 对象。

Python 示例如下。

在 PySpark 中的创建方法：进入 PySpark 以后就已经获得一个默认的 SparkConext 对象，即 sc。因此，我们可以采用以下方式来创建 StreamingContext 对象。

```
>>> from pyspark.streaming import StreamingContext
>>> ssc = StreamingContext(sc, 1)
```

如果只是编写一个独立的 Spark Streaming 程序，而不是在 PySpark 中运行，则需要通过以下方式创建 StreamingContext 对象。

```
from pyspark import SparkContext, SparkConf
from pyspark.streaming import StreamingContext
conf = SparkConf()
conf.setAppName('TestDStream')
conf.setMaster('local[2]')
sc = SparkContext(conf = conf)
ssc = StreamingContext(sc, 1)
```

在 Java 中，创建 'StreamingContext' 对象的步骤与 Python 类似，但语法略有不同。以下是 Java 的示例代码。

```
import org.apache.spark.SparkConf;
import org.apache.spark.streaming.StreamingContext;
import org.apache.spark.streaming.Durations;

public class StreamingApp {
    public static void main(String[] args) {
        SparkConf conf = new SparkConf().setAppName(
                        "TestDStream").setMaster("local[2]");
        StreamingContext ssc = new StreamingContext(conf,
                            Durations.seconds(1));

        // ··· 其他代码 ···

        ssc.start();
        ssc.awaitTermination();
    }
}
```

9.2.2 基本输入源

1．文件流

文件流即从文件中获取数据。

Python 示例如下。

```
#!/usr/bin/env python3
```

```
from pyspark import SparkContext, SparkConf
from pyspark.streaming import StreamingContext

conf = SparkConf()
conf.setAppName('TestDStream')
conf.setMaster('local[2]')
sc = SparkContext(conf = conf)
ssc = StreamingContext(sc, 10)
lines = ssc.textFileStream('file:///usr/local/spark/mycode/streaming/logfile')
words = lines.flatMap(lambda line: line.split(' '))
wordCounts = words.map(lambda x : (x,1)).reduceByKey(lambda a,b:a+b)
wordCounts.pprint()
ssc.start()
ssc.awaitTermination()
```

提交并运行 FileStreaming.py，代码如下。

```
$cd /usr/local/spark/mycode/streaming/logfile/
$/usr/local/spark/bin/spark-submit FileStreaming.py
```

Java 示例如下。

```
import org.apache.spark.SparkConf;
import org.apache.spark.api.java.JavaPairRDD;
import org.apache.spark.api.java.JavaRDD;
import org.apache.spark.api.java.function.FlatMapFunction;
import org.apache.spark.api.java.function.Function2;
import org.apache.spark.api.java.function.PairFunction;
import org.apache.spark.streaming.Duration;
import org.apache.spark.streaming.api.java.*;
import scala.Tuple2;
import java.util.Arrays;
import java.util.Iterator;

public class FileStreaming {
    public static void main(String[] args) {
        // 创建 SparkConf 对象
        SparkConf conf = new SparkConf().setAppName(
                        "TestDStream").setMaster("local[2]");

        // 创建 JavaStreamingContext 对象，第二个参数是批处理间隔
        JavaStreamingContext jssc = new JavaStreamingContext(conf,
                                    new Duration(10000));

        // 使用 textFileStream 创建 DStream
        JavaDStream<String> lines = jssc.textFileStream(
                        "file:///usr/local/spark/mycode/streaming/logfile");

        // 分割行为单词
        JavaDStream<String> words = lines.flatMap((FlatMapFunction<String,
```

```
        String>) line -> Arrays.asList(line.split(" ")).iterator());

        // 映射每个单词为一个键值对(key, value)，然后使用 reduceByKey 聚合结果
        JavaPairDStream<String, Integer> wordCounts = words.mapToPair(
                (PairFunction<String, String, Integer>) s -> new Tuple2<>
                (s, 1)).reduceByKey((Function2<Integer, Integer, Integer>) (a, b)
                -> a + b);

        // 打印结果
        wordCounts.print();

        // 开始流处理并等待终止
        jssc.start();
        try {
            jssc.awaitTermination();
        } catch (InterruptedException e) {
            e.printStackTrace();
        }
    }
}
```

上述代码可以在 IDEA 等开发工具中运行；也可以先编译并打包为 Java 代码，之后使用以下命令运行程序。

```
$cd /usr/local/spark/mycode/streaming/logfile/
$/usr/local/spark/bin/spark-submit --class YourJavaClassName -master
local[2] YourJarFileName.jar
```

2．套接字流

Spark Streaming 可以通过 Socket 端口监听并接收数据，然后进行相应的处理。

（1）Socket 工作原理

Spark Streaming 工作机制如图 9-12 所示。通信的基本步骤如下。

① 服务器端的基本步骤如下。

步骤 1：创建一个用于监听连接的 Socket 对象。

步骤 2：用指定的端口号和服务器的 IP 地址建立一个 EndPoint 对象。

步骤 3：用 Socket 对象的 bind()方法绑定 EndPoint。

步骤 4：用 Socket 对象的 listen()方法开始监听。

步骤 5：接收客户端的连接，用 Socket 对象的 accept()方法创建一个新的用于和客户端进行通信的 Socket 对象。

步骤 6：通信结束后一定要关闭 Socket。

② 客户端的基本步骤如下。

步骤 1：建立一个 Socket 对象。

步骤 2：用指定的端口号和服务器的 IP 地址建立一个 EndPoint 对象。

步骤 3：用 Socket 对象的 connect()方法以上面建立的 EndPoint 对象作为参数，向服务器发出连接请求。

步骤 4：如果连接成功，就用 Socket 对象的 send()方法向服务器发送信息。

步骤 5：用 Socket 对象的 receive()方法接收服务器发来的信息。

步骤 6：通信结束后一定要关闭 Socket。

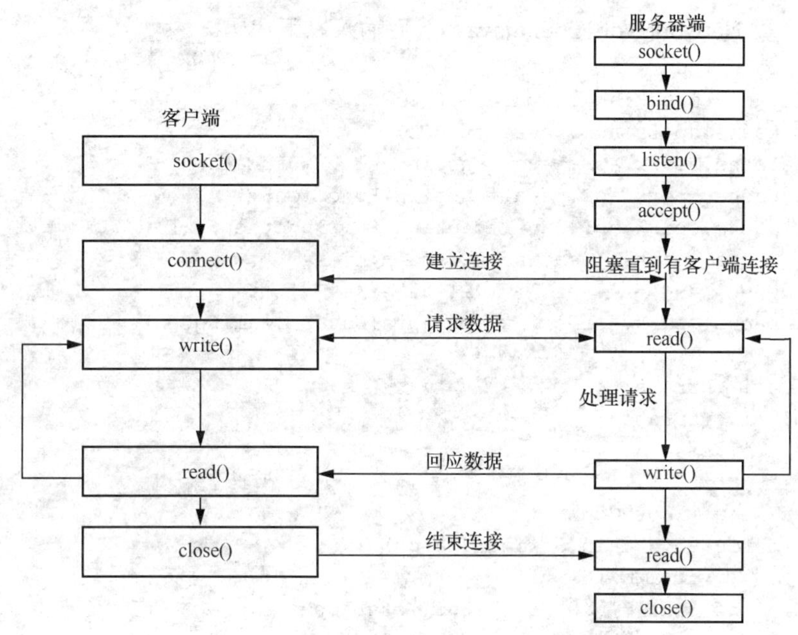

图 9-12 Spark Streaming 工作机制

（2）使用套接字流作为数据源

```
$cd /usr/local/spark/mycode
$mkdir streaming # 如果已经存在该目录，则不用创建
$cd streaming
$mkdir socket
$cd socket
```

Python 示例代码如下。

在名为 NetworkWordCount.py 文件中输入以下内容。

```
# !/usr/bin/env python3

from pyspark import SparkContext
from pyspark.streaming import StreamingContext

sc = SparkContext("local[2]", "NetworkWordCount")
ssc = StreamingContext(sc, 10)
lines = ssc.socketTextStream("localhost", 9999)
words = lines.flatMap(lambda line: line.split(""))
pairs = words.map(lambda word: (word, 1))
wordCounts = pairs.reduceByKey(lambda x, y: x + y)
wordCounts.pprint()
ssc.start()
ssc.awaitTermination()
```

再新建一个终端（记作“流计算终端”），执行以下代码启动流计算。

```
$cd /usr/local/spark/mycode/streaming/socket
$/usr/local/spark/bin/spark-submit NetworkWordCount.py localhost 9999
```

Java 示例代码如下。

新建一个类 NetworkWordCount.java，然后输入以下代码。

```
import org.apache.spark.SparkConf;
import org.apache.spark.api.java.JavaPairRDD;
import org.apache.spark.api.java.JavaRDD;
import org.apache.spark.api.java.function.FlatMapFunction;
import org.apache.spark.api.java.function.Function2;
import org.apache.spark.api.java.function.PairFunction;
import org.apache.spark.streaming.Duration;
import org.apache.spark.streaming.api.java.*;
import org.apache.spark.streaming.api.java.JavaStreamingContext;
import scala.Tuple2;
import java.util.Arrays;
import java.util.Iterator;
import java.util.List;

public class NetworkWordCount {
    public static void main(String[] args) {
        // 创建 SparkConf 对象
        SparkConf conf = new SparkConf().setAppName(
                "NetworkWordCount").setMaster("local[2]");

        // 创建 JavaStreamingContext 对象，第二个参数是批处理间隔
        JavaStreamingContext jssc = new JavaStreamingContext(conf,
                                    new Duration(10000));

        // 使用 socketTextStream 创建 DStream
        JavaReceiverInputDStream<String> lines = jssc.socketTextStream(
                                                "localhost", 9999);

        // 分割行为单词
        JavaDStream<String> words = lines.flatMap((FlatMapFunction<String,
        String>) line -> Arrays.asList(line.split("")).iterator());

        // 映射每个单词为一个(key, value)对，然后使用 reduceByKey 聚合结果
        JavaPairDStream<String, Integer> wordCounts = words.mapToPair(
            (PairFunction<String, String, Integer>) s -> new Tuple2<>(s, 1))
            .reduceByKey((Function2<Integer, Integer, Integer>) (a, b) -> a + b);

        // 打印结果
        wordCounts.print();

        // 开始流计算处理并等待终止
        jssc.start();
        try {
            jssc.awaitTermination();
        } catch (InterruptedException e) {
```

```
            e.printStackTrace();
        }
    }
}
```

上述代码在 IDEA 等开发环境中运行；也可以编译并打包 Java 代码，之后使用以下命令运行。

```
$cd /usr/local/spark/mycode/streaming/socket
$/usr/local/spark/bin/spark-submit --class NetworkWordCount --master local[2]
YourJarFileName.jar
```

新打开一个窗口作为数字控制（Number Control，NC）窗口，启动 NC 程序，命令如下。

```
$nc -lk 9999
```

在 NC 窗口中随意输入一些单词，监听窗口会自动获得单词数据流信息，并每隔 1s 打印出词频统计信息。词频统计信息类似于以下的结果。

```
-------------------------------------------
Time: 2018-12-24 11:30:26
-------------------------------------------
('Spark', 1)
('love', 1)
('I', 1)
(spark,1)
```

（3）使用 Socket 编程实现自定义数据源

下面我们修改数据源的产生方式。这里不使用 NC 程序，而是采用自己编写的程序产生 Socket 数据源，代码如下。

```
$cd /usr/local/spark/mycode/streaming/socket
$vim DataSourceSocket.py
```

在 DataSourceSocket.py 文件中输入以下内容。

```
# !/usr/bin/env python3
import socket
# 生成 Socket 对象
server = socket.socket()
# 绑定 ip 和端口
server.bind(('localhost', 9999))
# 监听绑定的端口
server.listen(1)
while 1:
    # 为了方便识别，打印一个“我在等待”
    print("我在等待")
    # 这里用两个值接收，因为连接上之后使用的是客户端发来请求的这个实例，
    # 所以下面的传输要使用 conn 实例操作
    conn,addr = server.accept()
    # 打印连接成功
    print("Connect success! Connection is from %s " % addr[0])
    # 打印正在发送数据
    print('Sending data...')
    conn.send('I love hadoop I love spark hadoop is good spark is fast'.encode())
    conn.close()
```

```
    print('Connection is broken.')
```

执行以下命令启动 Socket 服务端。

```
$cd /usr/local/spark/mycode/streaming/socket
$/usr/local/spark/bin/spark-submit DataSourceSocket.py
```

下面新建一个终端（记作流计算终端），输入以下命令启动 NetworkWordCount 程序，即客户端。

```
$cd  /usr/local/spark/mycode/streaming/socket
$/usr/local/spark/bin/spark-submit NetworkWordCount.py localhost 9999
```

运行结果如下。

```
-------------------------------------------
Time: 2018-12-30 15:16:17
-------------------------------------------
('good', 1)
('hadoop', 2)
('is', 2)
('love', 2)
('spark', 2)
('I', 2)
('fast', 1)
```

3．RDD 队列流

在调试 Spark Streaming 应用程序时，我们可以使用 streamingContext.queueStream(queueOfRDD)创建基于 RDD 队列的 DStream。

新建一个 RDDQueueStream.py 文件，该文件的功能是每隔 1 s 创建一个 RDD，Spark Streaming 每隔 2 s 对数据进行处理。

Python 示例代码如下。

```
# !/usr/bin/env python3

import time
from pyspark import SparkContext
from pyspark.streaming import StreamingContext

if __name__ == "__main__":
    sc = SparkContext(appName = "PythonStreamingQueueStream")
    ssc = StreamingContext(sc, 2)
    # 创建一个队列，通过该队列把 RDD 推给一个 RDD 队列流
    rddQueue = []
    for i in range(5):
        rddQueue += [ssc.sparkContext.parallelize([j for j in
                    range(1, 1001)], 10)]
        time.sleep(1)
    # 创建一个 RDD 队列流
    inputStream = ssc.queueStream(rddQueue)
    mappedStream = inputStream.map(lambda x: (x % 10, 1))
    reducedStream = mappedStream.reduceByKey(lambda a, b: a + b)
    reducedStream.pprint()
    ssc.start()
```

```
    ssc.stop(stopSparkContext=True, stopGraceFully=True)
```

下面执行以下命令运行该程序。

```
$cd  /usr/local/spark/mycode/streaming/rddqueue
$/usr/local/spark/bin/spark-submit RDDQueueStream.py
```

Java 示例代码如下。

```
import org.apache.spark.SparkConf;
import org.apache.spark.api.java.JavaRDD;
import org.apache.spark.api.java.JavaPairRDD;
import org.apache.spark.streaming.Duration;
import org.apache.spark.streaming.api.java.*;
import org.apache.spark.streaming.api.java.JavaStreamingContext;
import scala.Tuple2;

import java.util.LinkedList;
import java.util.Queue;

public class RDDQueueStream {
    public static void main(String[] args) throws InterruptedException {
        // 创建 SparkConf 对象
        SparkConf conf = new SparkConf().setAppName(
                        "JavaStreamingQueueStream");
        JavaStreamingContext ssc = new JavaStreamingContext(conf,
                                   new Duration(2000));

        // 创建一个 RDD 队列
        Queue<JavaRDD<Integer>> rddQueue = new LinkedList<>();
        for (int i = 0; i < 5; i++) {
            JavaRDD<Integer> rdd = ssc.sparkContext().parallelize(
                                    range(1, 1001), 10);
            rddQueue.offer(rdd);
            Thread.sleep(1000);
        }

        // 创建一个 RDD 队列流
        JavaInputDStream<Integer> inputStream = ssc.queueStream(rddQueue);
        JavaPairDStream<Integer, Integer> mappedStream =
        inputStream.mapToPair(x -> new Tuple2<>(x % 10, 1));
        JavaPairDStream<Integer, Integer> reducedStream =
        mappedStream.reduceByKey(Integer::sum);
        reducedStream.print();

        ssc.start();
        ssc.awaitTermination();
    }

    public static LinkedList<Integer> range(int start, int end) {
        LinkedList<Integer> list = new LinkedList<>();
        for (int i = start; i < end; i++) {
            list.add(i);
```

```
        }
        return list;
    }
}
```

上述代码可以在 IDEA 等工具中运行；也可以先编译并打包为 Java 代码，并执行以下语句运行。

```
$cd /usr/local/spark/mycode/streaming/rddqueue
$/usr/local/spark/bin/spark-submit --class RDDQueueStream YourJarFileName.jar
```

运行结果如下。

```
-------------------------------------------

Time: 2018-12-31 15:42:15
-------------------------------------------
(0, 100)
(8, 100)
(2, 100)
(4, 100)
(6, 100)
(1, 100)
(3, 100)
(9, 100)
(5, 100)
(7, 100)
```

9.2.3 转换操作

1. DStream 无状态转换操作

无状态转化操作就是当前批次的数据处理不依赖之前批次的数据。以下是部分无状态转化操作。

① map(func)：对源 DStream 的每个元素采用 func()函数进行转换，得到一个新的 DStream。

② flatMap(func)：与 map 操作相似，但是每个输入项可被映射为 0 个或者多个输出项。

③ filter(func)：返回一个新的 DStream，该 DStream 仅包含源 DStream 中满足 func()函数的项。

④ repartition(numPartitions)：通过创建更多或者更少的分区来改变 DStream 的并行程度。

⑤ reduce(func)：利用 func()函数聚集源 DStream 中每个 RDD 的元素，返回一个包含单元素 RDDs 的新 DStream。

⑥ count()：统计源 DStream 中每个 RDD 的元素数量。

⑦ union(otherStream)：返回一个新的 DStream，其中包含源 DStream 和其他 DStream 的元素。

⑧ countByValue()：应用于元素类型为键的 DStream 上，返回一个键值对类型的新

DStream，每个键的值是原 DStream 中每个 RDD 中的出现次数。

⑨ reduceByKey(func, [numTasks])：当在一个由键值对组成的 DStream 上执行该操作时，该操作返回一个新的由键值对组成的 DStream，每个键的值均由给定的 reduce()函数（func()函数）聚集起来。

⑩ join(otherStream, [numTasks])：当应用于两个 DStream——一个包含(K,V)键值对，一个包含(K,W)键值对时，返回一个包含(K, (V, W))键值对的新 DStream，其中，K 表示键，V 和 W 均表示值。

⑪ cogroup(otherStream, [numTasks])：当应用于两个 DStream——一个包含(K,V)键值对，一个包含(K,W)键值对时，返回一个包含(K, Seq[V], Seq[W])的元组。

⑫ transform(func)：通过对源 DStream 的每个 RDD 应用 RDD-to-RDD 函数来创建一个新的 DStream。该操作支持在新的 DStream 中做任何 RDD 操作。

前文中套接字流部分介绍的词频统计，采用的就是无状态转换操作：每次只统计当前批次到达的单词的词频，不会进行累计之前批次到达的单词的词频。

2．DStream 有状态转换操作

有状态转换操作指当前批次的处理需要使用之前批次的数据或者中间结果。有状态转换包括基于滑动窗口的转换和追踪状态变化的转换。

（1）基于滑动窗口的转换操作。

基于滑动窗口的转换如图 9-13 所示。

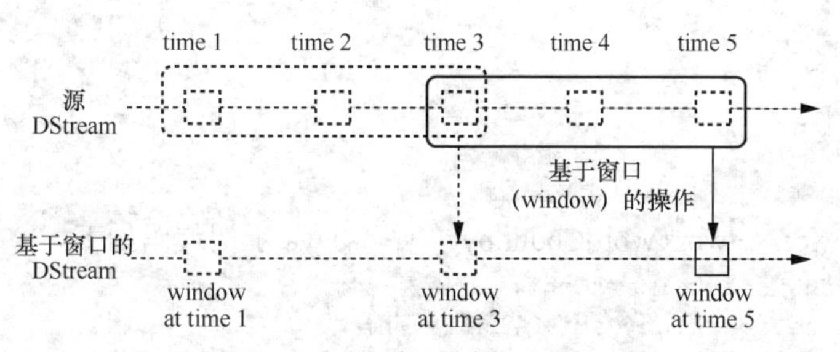

图 9-13　基于滑动窗口的转换

① 事先设定一个滑动窗口的长度（也就是窗口的持续时间）。

② 设定滑动窗口的时间间隔（每隔一段时间执行一次计算），让窗口按照指定时间间隔在源 DStream 上滑动。

③ 每次窗口停放的位置上都会有一部分 DStream（或者一部分 RDD）被框入窗口内，形成一个小段的 DStream。

④ 启动对这个小段 DStream 的计算。

一些窗口转换操作的含义如下。

Window(windowLength, slideInterval)：基于源 DStream 产生的窗口化的批数据，计算得到一个新的 DStream。

countByWindow(windowLength, slideInterval)：返回流中元素的一个滑动窗口数。

reduceByWindow(func, windowLength, slideInterval)：返回一个单元素流。利用 func()

函数聚集滑动时间间隔的流的元素来创建单元素流。func()函数必须满足结合律，以支持并行计算。

countByValueAndWindow(windowLength, slideInterval, [numTasks])：当应用到一个由键值对组成的 DStream 上时，该操作返回一个由键值对组成的新的 DStream。每个键的值是该键在滑动窗口中出现的频率。

reduceByKeyAndWindow(func, windowLength, slideInterval, [numTasks])：应用到一个由键值对组成的 DStream 上时，该操作会返回一个由键值对组成的新的 DStream。每一个键的值均由给定的 reduce()函数和 func()函数进行聚合计算。注意，在默认情况下，这个算子利用了 Spark 默认的并发任务数来分组。我们可以通过 numTasks 参数来指定任务数。

reduceByKeyAndWindow(func, invFunc, windowLength, slideInterval, [numTasks])：每个窗口的 reduce 值是基于先前窗口的 reduce 值进行增量计算得到的，该操作会对进入滑动窗口的新数据进行 reduce 操作，并对离开窗口的老数据进行"逆向 reduce"操作。

创建 WindowedNetworkWordCount.py 文件的 Python 示例代码如下。

```
from pyspark import SparkContext
from pyspark.streaming import StreamingContext
sc = SparkContext("local[2]", "NetworkWordCount")
ssc = StreamingContext(sc, 10)
ssc.checkpoint("file:///usr/local/spark/mycode/streaming/socket/checkpoint")
lines = ssc.socketTextStream("localhost", 9999)
counts = lines.flatMap(lambda line: line.split(""))\
.map(lambda word: (word, 1))\
.reduceByKeyAndWindow(lambda x, y: x + y, lambda x, y: x - y, 30, 10)
counts.pprint()
ssc.start()
ssc.awaitTermination()
```

创建 WindowedNetworkWordCount.py 文件的 Java 示例代码如下。

```
import org.apache.spark.SparkConf;
import org.apache.spark.streaming.Duration;
import org.apache.spark.streaming.api.java.*;
import org.apache.spark.streaming.api.java.JavaStreamingContext;
import scala.Tuple2;

import java.util.Arrays;
import java.util.List;

public class WindowedNetworkWordCount {
public static void main(String[] args) throws InterruptedException {
        // 设置 SparkConf
        SparkConf sparkConf = new SparkConf().setAppName(
                            "JavaNetworkWordCount").setMaster("local[2]");
        JavaStreamingContext ssc = new JavaStreamingContext(sparkConf,
                                  new Duration(10000));

        // 设置检查点
        ssc.checkpoint(
```

```
        "file:///usr/local/spark/mycode/streaming/socket/checkpoint");

        // 创建套接字文本流
        JavaReceiverInputDStream<String> lines = ssc.socketTextStream(
                                                  "localhost", 9999);

        // 将每行文本拆分为单词
        JavaDStream<String> words = lines.flatMap(x -> Arrays.asList(
                                    x.split("")).iterator());

        // 映射每个单词为(key, 1)并进行窗口化的 reduce 操作
        JavaPairDStream<String, Integer> wordCounts = words.mapToPair(s ->
        new Tuple2<>(s, 1)).reduceByKeyAndWindow((a, b) -> a + b, (a, b) -> a - b,
        new Duration(30000), new Duration(10000));

        wordCounts.print();

        ssc.start();
        ssc.awaitTermination();
    }
}
```

（2）追踪状态变化的转换

若需要在跨批次之间维护状态，则必须使用 updateStateByKey 操作。

词频统计实例如下。

对于有状态转换操作而言，本批次的词频统计会在之前批次的词频统计结果的基础上进行不断累加，所以，最终统计得到的词频是所有批次单词的词频统计结果。

创建 NetworkWordCountStateful.py 文件的 Python 示例代码如下。

```
from __future__ import print_function
import sys
from pyspark import SparkContext
from pyspark.streaming import StreamingContext
if __name__ == "__main__":
    sc = SparkContext("local[2]", "PythonStreamingStatefulNetworkWordCount")
    ssc = StreamingContext(sc, 1)
    ssc.checkpoint("file:///usr/local/spark/mycode/streaming/stateful/")
    # RDD with initial state (key, value) pairs
    initialStateRDD = sc.parallelize([(u'hello', 1), (u'world', 1)])
    def updateFunc(new_values, last_sum):
        return sum(new_values) + (last_sum or 0)
    lines = ssc.socketTextStream(sys.argv[1], int(sys.argv[2]))
    running_counts = lines.flatMap(lambda line: line.split(""))\
                          .map(lambda word: (word, 1))\
                          .updateStateByKey(updateFunc,
                                            initialRDD = initialStateRDD)
    running_counts.pprint()
    ssc.start()
ssc.awaitTermination()
```

新建一个 Linux 终端（记为流计算终端），执行以下命令提交运行程序。

```
$cd /usr/local/spark/mycode/streaming/stateful
$/usr/local/spark/bin/spark-submit \
> NetworkWordCountStateful.py localhost 9999
```

创建 NetworkWordCountStateful.py 文件的 Java 示例如下。

```
import org.apache.spark.SparkConf;
import org.apache.spark.api.java.JavaPairRDD;
import org.apache.spark.api.java.JavaRDD;
import org.apache.spark.streaming.Durations;
import org.apache.spark.streaming.api.java.*;
import scala.Tuple2;
import java.util.Arrays;
import java.util.List;
import java.util.Optional;

public class NetworkWordCountStateful {

    private static Optional<Integer> updateFunc(List<Integer> newValues,
    Optional<Integer> runningCount) {
        Integer newCount = runningCount.orElse(0) +
                           newValues.stream().mapToInt(Integer::intValue).sum();
        return Optional.of(newCount);
    }

    public static void main(String[] args) {
        SparkConf conf = new SparkConf().setMaster("local[2]").setAppName(
        "JavaNetworkWordCountStateful");
        JavaStreamingContext jssc = new JavaStreamingContext(conf,
                                    Durations.seconds(1));

        jssc.checkpoint("file:///usr/local/spark/mycode/streaming/stateful/");

        // Initial state RDD input to mapWithState
        List<Tuple2<String, Integer>> tuples = Arrays.asList(new Tuple2<>(
        "hello", 1), new Tuple2<>("world", 1));
        JavaPairRDD<String, Integer> initialRDD =
        jssc.sparkContext().parallelizePairs(tuples);

        JavaReceiverInputDStream<String> lines = jssc.socketTextStream(
        args[0], Integer.parseInt(args[1]));

        JavaDStream<String> words = lines.flatMap(x -> Arrays.asList(
                                    x.split(" ")).iterator());

        JavaPairDStream<String, Integer> wordDstream = words.mapToPair(
        word -> new Tuple2<>(word, 1));

        JavaPairDStream<String, Integer> runningCounts =
        wordDstream.updateStateByKey(NetworkWordCountStateful::updateFunc,
```

```
        initialRDD);

        runningCounts.print();

        jssc.start();
        try {
            jssc.awaitTermination();
        } catch (InterruptedException e) {
            e.printStackTrace();
        }
    }
}
```

以上代码可以在 IDEA 等开发工具中运行；也可以编译并打包为 Java 程序后，用以下命令运行。

```
$cd /usr/local/spark/mycode/streaming/stateful
$/usr/local/spark/bin/spark-submit --class NetworkWordCountStateful
YourJavaJarName.jar localhost 9999
```

新建一个终端（记为数据源终端），执行以下命令启动 NC 程序。

```
$nc  -lk  9999
```

在数据源终端内手动输入一些单词并按回车键，再切换到流计算终端，这时可以看到输出了类似以下内容的词频统计信息。

```
-------------------------------------------
Time: 2018-12-30 20:53:02
-------------------------------------------
('world', 1)
('hello', 1)
```

9.2.4 输出操作

在 Spark 应用中，外部系统经常需要使用 Spark DStream 处理后的数据，因此，我们需要采用输出操作把 DStream 的数据输出到数据库或者文件系统中。

1. 把 DStream 输出到文本文件中

Python 示例代码如下。在 NetworkWordCountStatefulText.py 代码文件中输入以下内容。

```
import sys
from pyspark import SparkContext
from pyspark.streaming import StreamingContext
if __name__ == "__main__":
    sc = SparkContext("local[2]", "PythonStreamingStatefulNetworkWordCount")
    ssc = StreamingContext(sc, 1)
    ssc.checkpoint("file:///usr/local/spark/mycode/streaming/stateful/")
    # RDD with initial state (key, value) pairs
    initialStateRDD = sc.parallelize([(u'hello', 1), (u'world', 1)])
    def updateFunc(new_values, last_sum):
        return sum(new_values) + (last_sum or 0)
    lines = ssc.socketTextStream(sys.argv[1], int(sys.argv[2]))
```

```
    running_counts = lines.flatMap(lambda line: line.split(""))\
                          .map(lambda word: (word, 1))\
                          .updateStateByKey(updateFunc, initialRDD =
                                            initialStateRDD)
    running_counts.saveAsTextFiles(
    "file:///usr/local/spark/mycode/streaming/stateful/output")
    running_counts.pprint()
    ssc.start()
    ssc.awaitTermination()
```

新建 NetworkWordCountStatefulText.java 文件，Java 示例代码如下。

```
import org.apache.spark.SparkConf;
import org.apache.spark.api.java.JavaPairRDD;
import org.apache.spark.api.java.JavaRDD;
import org.apache.spark.streaming.Durations;
import org.apache.spark.streaming.api.java.*;
import scala.Tuple2;
import java.util.Arrays;
import java.util.List;
import java.util.Optional;

public class NetworkWordCountStatefulText {

    private static Optional<Integer> updateFunc(List<Integer> newValues,
    Optional<Integer> runningCount) {
        Integer newCount = runningCount.orElse(0) +
                           newValues.stream().mapToInt(Integer::intValue).sum();
        return Optional.of(newCount);
    }

    public static void main(String[] args) {
        if (args.length < 2) {
            System.err.println("Usage: NetworkWordCountStatefulText
                                   <hostname> <port>");
            System.exit(1);
        }

        SparkConf conf = new SparkConf().setMaster("local[2]").setAppName(
                    "JavaNetworkWordCountStatefulText");
        JavaStreamingContext jssc = new JavaStreamingContext(conf,
                                      Durations.seconds(1));

        jssc.checkpoint("file:///usr/local/spark/mycode/streaming/stateful/");

        // Initial state RDD input
        List<Tuple2<String, Integer>> tuples = Arrays.asList(new Tuple2<>(
        "hello", 1), new Tuple2<>("world", 1));
        JavaPairRDD<String, Integer> initialRDD =
        jssc.sparkContext().parallelizePairs(tuples);
```

```
        JavaReceiverInputDStream<String> lines = jssc.socketTextStream(
                                        args[0], Integer.parseInt(args[1]));

        JavaDStream<String> words =
        lines.flatMap(x -> Arrays.asList(x.split("")).iterator());

        JavaPairDStream<String, Integer> wordDstream =
        words.mapToPair(word -> new Tuple2<>(word, 1));

        JavaPairDStream<String, Integer> runningCounts =
        wordDstream.updateStateByKey(NetworkWordCountStatefulText::updateFunc,
        initialRDD);

        runningCounts.saveAsTextFiles(
        "file:///usr/local/spark/mycode/streaming/stateful/output");
        runningCounts.print();

        jssc.start();
        try {
            jssc.awaitTermination();
        } catch (InterruptedException e) {
            e.printStackTrace();
        }
    }
}
```

2．把 DStream 写入关系数据库

启动 MySQL 数据库，并完成数据库和数据表的创建。

```
$service mysql start
$mysql -u root -p
```

这时系统会提示你输入密码，你按要求输入即可。

在此前已经创建好的 spark 数据库中创建一个名称为 wordcount 的数据表，命令如下。

```
mysql> use spark
mysql> create table wordcount (word char(20), count int(4));
```

由于需要让 Python 连接 MySQL 数据库，所以，需要先安装 Python 连接 MySQL 的模块 PyMySQL。我们在 Linux 终端中执行以下命令。

```
$sudo apt-get update
$sudo apt-get install python3-pip
$pip3 -V
$sudo pip3 install PyMySQL
```

添加 NetworkWordCountStatefulDB.py 文件，Python 示例代码如下。

```
from __future__ import print_function
import sys
import pymysql
from pyspark import SparkContext
from pyspark.streaming import StreamingContext
```

```
if __name__ == "__main__":
    sc = SparkContext("local[2]","PythonStreamingStatefulNetworkWordCount")
    ssc = StreamingContext(sc, 1)
    ssc.checkpoint("file:///usr/local/spark/mycode/streaming/stateful")
    # RDD with initial state (key, value) pairs
    initialStateRDD = sc.parallelize([(u'hello', 1), (u'world', 1)])

    def updateFunc(new_values, last_sum):
        return sum(new_values) + (last_sum or 0)

    lines = ssc.socketTextStream(sys.argv[1], int(sys.argv[2]))
    running_counts = lines.flatMap(lambda line: line.split(""))\
                          .map(lambda word: (word, 1))\
                          .updateStateByKey(updateFunc, initialRDD =
                                            initialStateRDD)
    running_counts.pprint()
def dbfunc(records):
        db = pymysql.connect("localhost","root","123456","spark")
        cursor = db.cursor()
        def doinsert(p):
            sql = "insert into wordcount(word,count) values ('%s', '%s')" % (
                  str(p[0]), str(p[1]))
            try:
                cursor.execute(sql)
                db.commit()
            except:
                db.rollback()
        for item in records:
            doinsert(item)
 def func(rdd):
        repartitionedRDD = rdd.repartition(3)
        repartitionedRDD.foreachPartition(dbfunc)

    running_counts.foreachRDD(func)
    ssc.start()
    ssc.awaitTermination()
```

新建 NetworkWordCountStatefulDB.java 文件，Java 示例代码如下。

```
import org.apache.spark.SparkConf;
import org.apache.spark.api.java.JavaPairRDD;
import org.apache.spark.api.java.JavaRDD;
import org.apache.spark.streaming.Durations;
import org.apache.spark.streaming.api.java.*;
import scala.Tuple2;

import java.sql.Connection;
import java.sql.DriverManager;
import java.sql.PreparedStatement;
import java.util.Arrays;
import java.util.List;
```

```
import java.util.Optional;

public class NetworkWordCountStatefulDB {

    private static Optional<Integer> updateFunc(List<Integer> newValues,
    Optional<Integer> runningCount) {
        Integer newCount = runningCount.orElse(0) +
                       newValues.stream().mapToInt(Integer::intValue).sum();
        return Optional.of(newCount);
    }

    private static void dbFunc(Iterator<Tuple2<String, Integer>> partition
    OfRecords) {
        String url = "jdbc:mysql://localhost/spark";
        String user = "root";
        String password = "123456";
        try (Connection connection = DriverManager.getConnection(url,
        user, password)) {
            while (partitionOfRecords.hasNext()) {
                Tuple2<String, Integer> tuple = partitionOfRecords.next();
                String sql = "INSERT INTO wordcount (word, count) VALUES (?, ?)";
                try (PreparedStatement statement =
                    connection.prepareStatement(sql)) {
                    statement.setString(1, tuple._1);
                    statement.setInt(2, tuple._2);
                    statement.executeUpdate();
                }
            }
        } catch (Exception e) {
            e.printStackTrace();
        }
    }

    public static void main(String[] args) {
        if (args.length < 2) {
            System.err.println("Usage:
            NetworkWordCountStatefulDB <hostname> <port>");
            System.exit(1);
        }

        SparkConf conf =
        new SparkConf().setMaster("local[2]").setAppName(
                                          "JavaNetworkWordCountStatefulDB");
        JavaStreamingContext jssc = new JavaStreamingContext(conf,
                                  Durations.seconds(1));

        jssc.checkpoint("file:///usr/local/spark/mycode/streaming/stateful/");
```

```
        List<Tuple2<String, Integer>> tuples =
        Arrays.asList(new Tuple2<>("hello", 1), new Tuple2<>("world", 1));
        JavaPairRDD<String, Integer> initialRDD =
        jssc.sparkContext().parallelizePairs(tuples);

        JavaReceiverInputDStream<String> lines =
        jssc.socketTextStream(args[0], Integer.parseInt(args[1]));

        JavaDStream<String> words =
        lines.flatMap(x -> Arrays.asList(x.split("")).iterator());

        JavaPairDStream<String, Integer> wordDstream =
        words.mapToPair(word -> new Tuple2<>(word, 1));

        JavaPairDStream<String, Integer> runningCounts =
        wordDstream.updateStateByKey(NetworkWordCountStatefulDB::updateFunc,
                                   initialRDD);

        runningCounts.print();

        runningCounts.foreachRDD(rdd -> {
            rdd.foreachPartition(NetworkWordCountStatefulDB::dbFunc);
        });

        jssc.start();
        try {
            jssc.awaitTermination();
        } catch (InterruptedException e) {
            e.printStackTrace();
        }
    }
}
```

注意：如果你的项目类型是 Maven 项目，请确保在 pom.xml 或类似的构建文件中包含以下依赖。

```
<dependency>
    <groupId>mysql</groupId>
    <artifactId>mysql-connector-java</artifactId>
    <version>8.0.23</version> <!-- Replace with your version -->
</dependency>
```

9.3 编程实现——流数据过滤与分析

【任务描述】运用实例，了解 Spark Streaming 结合 SparkSQL 的方法，掌握传输数据、查询、计算结果等操作。

采用 NC 程序传入数据，命令如下。

```
$nc  -lk  9999
classId name age sex  course score
12      张小军  25  男 chinese   50
12      张小军  25  男    math   60
12      张小军  25  男 english   70
12      李小凤  20  女 chinese   80
12      李小凤  20  女    math   80
12      李小凤  20  女 english   80
12      王大力  19  男 chinese   70
12      王大力  19  男    math   80
12      王大力  19  男 english   90
13      张大明  25  男 chinese   55
13      张大明  25  男    math   65
13      张大明  25  男 english   60
13      李小华  20  男 chinese   95
13      李小华  20  男    math   92
13      李小华  20  男 english   91
13      王小芳  19  女 chinese   75
13      王小芳  19  女    math   85
13      王小芳  19  女 english   90
```

分别用 Python 和 Java 实现以下功能。

① 输出采用的是 foreachRDD()算子。

② 将 RDD 转为 DataFrame 并输出，输出内容如图 9-14 所示。

③ 查询参加考试的人名。

④ 查询参加考试的男生姓名。

⑤ 输出参加考试中年龄等于 20 岁的人名。

⑥ 统计年龄大于 20 岁的人数。

⑦ 12 班有哪些人参加考试。

⑧ 计算语文科目的平均成绩。

⑨ 统计所有人中总成绩大于 150 分的人，输出他们的名字及总成绩。

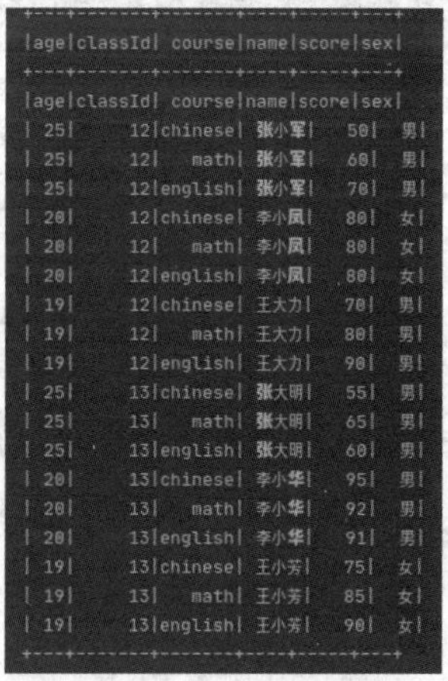

```
+---+-------+-------+---+-----+---+
|age|classId| course|name|score|sex|
+---+-------+-------+---+-----+---+
|age|classId| course|name|score|sex|
| 25|     12|chinese| 张小军|   50|  男|
| 25|     12|   math| 张小军|   60|  男|
| 25|     12|english| 张小军|   70|  男|
| 20|     12|chinese| 李小凤|   80|  女|
| 20|     12|   math| 李小凤|   80|  女|
| 20|     12|english| 李小凤|   80|  女|
| 19|     12|chinese| 王大力|   70|  男|
| 19|     12|   math| 王大力|   80|  男|
| 19|     12|english| 王大力|   90|  男|
| 25|     13|chinese| 张大明|   55|  男|
| 25|     13|   math| 张大明|   65|  男|
| 25|     13|english| 张大明|   60|  男|
| 20|     13|chinese| 李小华|   95|  男|
| 20|     13|   math| 李小华|   92|  男|
| 20|     13|english| 李小华|   91|  男|
| 19|     13|chinese| 王小芳|   75|  女|
| 19|     13|   math| 王小芳|   85|  女|
| 19|     13|english| 王小芳|   90|  女|
+---+-------+-------+---+-----+---+
```

图 9-14　输出内容

Python 示例代码如下。

```
# SparkStreamingToSparkSQL.py
from pyspark.sql import SparkSession
from pyspark.sql import Row
from pyspark import SparkContext
from pyspark.streaming import StreamingContext

sc = SparkContext("local[2]","app")
ssc = StreamingContext(sc, 10)
# ssc.checkpoint("checkpoint")
sqlContext = SparkSession.builder.getOrCreate()
```

```
lines = ssc.socketTextStream("localhost",1234)
students = lines.map(lambda line : line.split(""))

def getSparkSessionInstance(sparkConf):
    if("sparkSessionSingletonInstance" not in globals()):
        globals()['sparkSessionSingletonInstance'] =
        SparkSession.builder.getOrCreate()
        return globals()["sparkSessionSingletonInstance"]
def f1(rdd):
    len = rdd.count()
    if len > 0:
        spark = getSparkSessionInstance(rdd.context.getConf())
        rowRdd = rdd.map(lambda x: Row(classId = x[0], name = x[1],
                         age = x[2], sex = x[3], course = x[4], score = x[5]))
        print(rowRdd.take(5))
        studentsDataFrame = spark.createDataFrame(rowRdd)
        studentsDataFrame.createOrReplaceTempView("students")
        # 输出 DataFrame
        spark.sql("select * from students").show()
        # 查询参加考试人名
        spark.sql("select distinct name from students").show()
        # 查询参加考试的男生姓名
        spark.sql("select distinct name from students where sex ='男'").show()
        # 输出参加考试中年龄等于 20 岁的人名
        spark.sql("select distinct name from students where age = '20'").show()
        # 统计年龄大于 20 岁的人数
        spark.sql("select count(distinct name) from students where age> =
                '20'").show()
        # 12 班有哪些人参加考试
        spark.sql("select distinct name from students where classId = '12'").show()
        # 计算语文科目的平均成绩
        spark.sql("select avg(score) from students where course = 'chinese'").show()
        # 统计所有人中总成绩大于 150 分的人，输出他们的名字及总成绩
        spark.sql("select name,sum(score) from students group by name order
        by sum(score)>150").show()

students.foreachRDD(f1) # 采用 foreachRDD()算子输出
ssc.start()  # 开始计算过程
ssc.awaitTermination(120)  # 等待计算结束
ssc.stop(True, False)
```

Java 示例代码如下。

```
import org.apache.spark.SparkConf;
import org.apache.spark.api.java.JavaRDD;
import org.apache.spark.api.java.function.Function;
import org.apache.spark.api.java.function.VoidFunction;
import org.apache.spark.sql.Dataset;
import org.apache.spark.sql.Row;
import org.apache.spark.sql.SparkSession;
import org.apache.spark.streaming.Duration;
```

```
import org.apache.spark.streaming.api.java.JavaReceiverInputDStream;
import org.apache.spark.streaming.api.java.JavaStreamingContext;

public class SparkStreamingExample {

    public static void main(String[] args) {

        SparkConf conf = new SparkConf().setAppName("app").setMaster("local[2]");
        JavaStreamingContext ssc = new JavaStreamingContext(conf,
                                new Duration(10000));

        JavaReceiverInputDStream<String> lines = ssc.socketTextStream(
                                                "localhost", 1234);

        lines.foreachRDD(new VoidFunction<JavaRDD<String>>() {
            @Override
            public void call(JavaRDD<String> rdd) throws Exception {
                SparkSession spark =
                SparkSession.builder().config(
                                       rdd.context().getConf()).getOrCre ate();
                JavaRDD<Row> rowRdd = rdd.map(new Function<String, Row>() {
                    @Override
                    public Row call(String line) {
                        String[] parts = line.split("");
                        return RowFactory.create(parts[0], parts[1],
                        parts[2], parts[3], parts[4], parts[5]);
                    }
                });

                // Define the schema for the DataFrame
                StructType schema = DataTypes.createStructType(
                                    new StructField[]{
                    DataTypes.createStructField(
                    "classId", DataTypes.StringType, true),
                    DataTypes.createStructField(
                    "name", DataTypes.StringType, true),
                    DataTypes.createStructField(
                    "age", DataTypes.StringType, true),
                    DataTypes.createStructField(
                    "sex", DataTypes.StringType, true),
                    DataTypes.createStructField(
                    "course", DataTypes.StringType, true),
                    DataTypes.createStructField(
                    "score", DataTypes.StringType, true)
                });

                Dataset<Row> studentsDataFrame =
                spark.createDataFrame(rowRdd, schema);
                studentsDataFrame.createOrReplaceTempView("students");
```

```
                spark.sql("select * from students").show();
                spark.sql("select distinct name from students").show();
                spark.sql("select distinct name from students where sex =
                        '男'").show();
                spark.sql("select distinct name from students where age =
                        '20'").show();
                spark.sql("select count(distinct name) from students where
                        age>='20'").show();
                spark.sql("select distinct name from students where classId =
                        '12'").show();
                spark.sql("select avg(score) from students where course =
                        'chinese'").show();
                spark.sql("select name,sum(score) from students group by
                        name order by sum(score)>150").show();
            }
        });

        ssc.start();
        try {
            ssc.awaitTermination(120000); // 2 minutes
        } catch (InterruptedException e) {
            e.printStackTrace();
        }
        ssc.stop(true, false);
    }
}
```

9.4 本章小结

本章详细探讨了 Spark Streaming 这一实时计算框架。首先，本章对 Spark Streaming 的基本概念及其运行原理进行了系统性的阐述。其次，本章介绍了 Spark Streaming 的基础应用方法，为读者日后的编程实践奠定了坚实的基础。再次，本章深入到 DStream 编程模型层面，详细介绍转换操作、窗口操作及输出操作等内容。最后，本章通过一个结合 Spark Streaming 与 SparkSQL 的实例，使读者对 Spark Streaming 的运用有更深入的理解。

第 10 章 Spark GraphFrames：图计算

GraphX 是 Spark 中用于图和图计算的组件，GraphX 通过扩展 Spark RDD 引入了一个新的图抽象数据结构——一个将有效信息放入顶点和边的有向多重图。和 Spark 的每个组件一样，GraphX 有一个基于 RDD 的便于自己计算的抽象数据结构（例如 SQL 的 DataFrame、Streaming 的 DStream）。为了便于图计算，GraphX 公开了一系列基本运算（InDegress、OutDegress、subgraph 等），也有许多用于图计算算法工具包（PageRank、TriangleCount、ConnectedComponents 等）。

相对于其他分布式图计算框架，GraphX 最大的贡献（也是大多数开发人员喜欢它的原因）是在 Spark 之上提供了“一站式”解决方案，可以方便且高效地完成图计算的一整套流水作业。在实际开发中，用户可以使用核心模块来完成海量数据的清洗与分析，使用 SQL 模块打通与数据仓库的通道，使用 Streaming 打造实时流计算处理通道，基于 GraphX 图计算算法来对网页中复杂的业务关系进行计算，使用 MLLib 及 SparkR 来完成数据挖掘算法处理。

【学习目标】

1．认识并了解 Spark GraphFrames。
2．掌握 GraphFrames 编程模型的相关内容。
3．理解基于 GraphFrames 网页排名思路。

10.1 图计算概述

【任务描述】掌握图计算的方法，完成信任网络分析的任务。

10.1.1 图的基本概念

这里的图不是图片的意思，而是一种数据结构，该结构由一个有穷非空顶点集合和一个描述顶点间多对多关系的边集合组成。图的结构通常表示为 $G(V, E)$，其中，G 表示一个图，V 表示图 G 中顶点的集合，E 表示图 G 中边的集合。

图是一种复杂的非线性结构，在其结构中，每个元素都可以有 0 个或多个前驱，也可以有 0 个或多个后继。也就是说，在图中，元素之间的关系是任意的。每个图的顶点代表一个重要的对象，每条边则代表两个对象之间的关系。

图可以按照无方向和有方向分为无向图和有向图。如果任意两个顶点之间都存在边，则该图叫作完全图。有向的边叫作有向完全图。如果是无重复的边或者顶点到自身的边，则该图叫作简单图。有向图和无向图如图 10-1 所示。

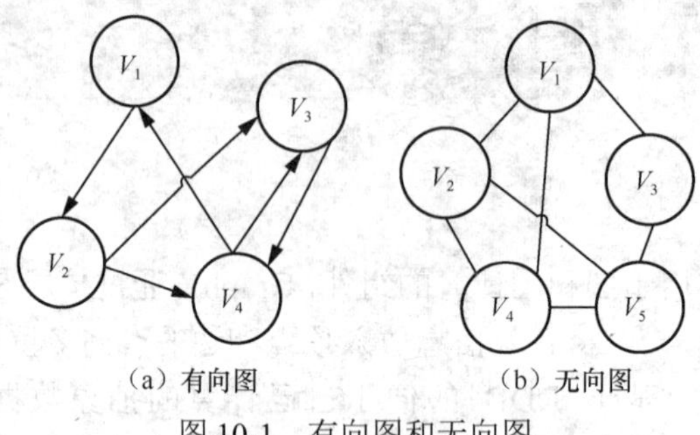

图 10-1　有向图和无向图

10.1.2　图计算的应用

图计算是一种以图论为基础，对现实世界的一种图结构的抽象表达，以及在这种数据结构上的计算模式。图结构很好地表达了数据之间的关联性，而关联性计算是大数据计算的核心。通过获得数据的关联性，我们可以从包括噪声的海量数据中抽取有用的信息。图计算的应用案例如下。

1．图谱计算平台

将各种图的指标精细化和规范化，为产品和运营的构思进行数据上的预研指导提供科学决策依据，这是图谱计算平台设计的初衷和出发点。基于这样的出发点，淘宝公司借助 GraphX 丰富的接口和工具包，针对淘宝公司内部林林总总的业务需求，开发了一个图谱计算平台。

2．PageRank 网页排名

PageRank 通过网络浩瀚的超链接关系来确定一个页面的等级。Google 公司把从 A 页面到 B 页面的链接解释为 A 页面给 B 页面投票，根据投票来源（甚至来源的来源，即链接到 A 页面的页面）和投票目标的等级决定新的等级。PageRank 通过网页之间的连接网络图来计算网页等级，是 Google 公司网页排名中的重要算法。

3．社交网络分析

社交网络本身就是一个复杂的图关系结构的网络，最适合用图来表达和计算：图的顶点表示社交中的人，边表示人与人之间的关系。例如，新浪微博社交网络分析通过用户之间的关注、转发等行为，建立了用户之间的社交网络关系图，根据用户在社交网络中所占的位置为用户进行分析和应用。

4．推荐应用

像淘宝网推荐商品、微信推荐好友等应用是将商品之间的交互做成一张大的网络图，将用户之间的关系做成一张社交网络图。在应用过程中，它们可以通过点与点之间的关系将与某商品相关的其他商品推荐给用户，或将朋友的朋友推荐给你。

10.1.3 GraphFrames 简介

GraphFrames 构建在 Spark DataFrames 之上，既能利用 DataFrame 良好的扩展性和强大的性能，同时也为 Scala、Java 和 Python 提供统一的图处理 API。GraphX 基于 RDD API，不支持 Python API。但是，GraphFrame 基于 DataFrame，并且支持 Python API。

目前 GraphFrames 还未集成到 Spark 中，只是作为单独的项目存在。GraphFrames 遵循与 Spark 相同的代码质量标准，并且是针对大量 Spark 版本进行交叉编译和发布的。

与 Apache Spark 的 GraphX 类似，GraphFrames 支持多种图处理功能，具有以下优势。

① 统一的 API：为 Python、Java 和 Scala 这 3 种语言提供了统一的接口，这是 Python 和 Java 能够使用 GraphX 的全部算法的原因。

② 强大的查询功能：GraphFrames 使用户可以编写与 Spark SQL 及 DataFrame 类似的查询语句。

③ 图的存储和读取：GraphFrames 与 DataFrame 的数据源完全兼容，支持以 Parquet、JSON、CSV 等格式完成图的存储或读取。在 GraphFrames 中，图的顶点和边都是以 DataFrame 形式存储的，所以一个图的所有信息都能够被完整保存。

④ GraphFrames 可以实现与 GraphX 的完美集成，它们两者之间在进行相互转换时不会丢失任何数据。

10.2 GraphFrames 编程模型基础

【任务描述】掌握 GraphFrame 的实例化图的创建、视图和图操作、图加载和保存等方法。

10.2.1 创建实例化图

GraphFrame 是 GraphFrames API 的核心抽象编程模型，是图的抽象。顶点 DataFrame 必须包含列名为 id 的列，作为顶点的唯一标识 id 表示顶点 V 队列 $G\in<V, E>$；边 DataFrame 必须包含列名为 src 和 dst 的列，保存头和尾的唯一标识 id。实例代码如下。

```
>>> from graphframes import *
>>> vertices = spark.createDataFrame([ ("a", "Alice", 34),("b", "Bob", 36)]  ,
["id", "name", "age"])
>>> edges = spark.createDataFrame([ ("a", "b", "friend")]  , ["src", "dst",
"relationship"])
>>> from graphframes import GraphFrame
>>> graph = GraphFrame(vertices,edges)
```

10.2.2 视图和图操作

GraphFrame 提供了 4 种视图，它们的返回类型都是 DataFrame。

① 顶点表视图：graph.Vertices.show()。

② 边表视图：graph.edges.show()。

③ 三元组（Triplet）视图：graph.triplets.show()。

④ 模式（Pattern）视图。

前 3 种视图的对应代码如下。

```
# 顶点表视图
>>> graph.vertices.show()
+---+-----+---+
| id| name|age|
+---+-----+---+
|  a|Alice| 34|
|  b|  Bob| 36|
+---+-----+---+
# 边表视图
>>> graph.edges.show()
+---+---+------------+
|src|dst|relationship|
+---+---+------------+
|  a|  b|      friend|
+---+---+------------+
# 三元组（Triplet）视图
>>> graph.triplets.show()
+--------------+--------------+-----------+
|           src|          edge|        dst|
+--------------+--------------+-----------+
|[a, Alice, 34]|[a, b, friend]|[b, Bob, 36]|
+--------------+--------------+-----------+
```

通过 GraphFrame 提供的 3 个属性 degrees、inDegrees、 outDegrees，我们可以获得顶点的度、入度和出度。

代码如下。

```
# degrees，顶点的度
>>> graph.degrees.show()
+---+------+
| id|degree|
+---+------+
|  b|     1|
|  a|     1|
+---+------+
# inDegrees，顶点的入度
>> graph.inDegrees.show()
+---+--------+
| id|inDegree|
```

```
+---+--------+
|  b|       1|
+---+--------+
# outDegrees，顶点的出度
>> graph.outDegrees.show()
+---+---------+
| id|outDegree|
+---+---------+
|  a|        1|
+---+---------+
```

除了上述 3 种基本视图，GraphFrame 通过 find()方法提供了与 Neo4j 的 Cypher 查询类似的模式查询功能，其返回的 DataFrame 类型的结果被称为模式视图。模式视图使用一种简单的领域专用语言（Domain Specific Language，DSL）语言来实现图的结构化查询，采用形如“(a)-[e]->(b)”的模式来描述一条有向边，其中，(a)、(b)表示顶点，a 和 b 表示顶点名，[e]表示边，e 表示边名，->表示有向边的方向。顶点名和边名会作为搜索结果的列名，如果结果中不需要该项，可在模式中省略该名称。另外，当模式中有多条边时，需要用分号“;”拼接，例如，“(a)-[e]->(b); (b)-[e2]->(c)”表示一条先从 a 到 b，然后从 b 到 c 的边。如果不包含某条边，则可以在表示边的模式前面加上“!”，例如，“(a)-[e]->(b); !(b)-[e2]->(a)”，表示不选取包含重边的边。示例代码如下。

```
>>>motifis = graph.find(“(a)-[e]->(b)”)
>>>motifis.show()
```

graph.find()的运行结果如图 10-2 所示。

```
>>> motifis = graph.find("(a)-[e]->(b)")
>>> motifis.show()
+--------------+---------------+------------+
|             a|              e|           b|
+--------------+---------------+------------+
|[a, Alice, 34]|[a, b, friend]|[b, Bob, 36]|
+--------------+---------------+------------+
```

图 10-2　graph.find()的运行结果

模式视图属于 DataFrame 类型，同样可以进一步进行查询、过滤和统计操作，代码如下。

```
>>>motifis.filter(“b.age>30”).show()
```

motifis.filter()的运行结果如图 10-3 所示。

```
>>> motifis.filter("b.age>30").show()
+--------------+---------------+------------+
|             a|              e|           b|
+--------------+---------------+------------+
|[a, Alice, 34]|[a, b, friend]|[b, Bob, 36]|
+--------------+---------------+------------+
```

图 10-3　motifis.filter()的运行结果

10.2.3 保存和加载图

1. 保存图

图可以以多种格式来保存，这里介绍 parquet 和 json 格式的图保存。

Parquet 格式的图保存示例代码如下。

```
# 注意这个保存路径属于 HDFS
>>>graph.vertices.write.parquet("hdfs:// xdata-m1:8020/user/ua50/vertices")
>>>graph.edges.write.parquet("hdfs:// xdata-m1:8020/user/ua50/edges")
```

Parquet 格式保存的运行结果如图 10-4 所示。

```
hadoop@master:~/spark-graphX$ ls -A
edges  vertices
hadoop@master:~/spark-graphX$ ls -l
total 8
drwxrwxr-x 2 hadoop hadoop 4096 May 12 03:42 edges
drwxrwxr-x 2 hadoop hadoop 4096 May 12 03:41 vertices
hadoop@master:~/spark-graphX$ cd edges/
hadoop@master:~/spark-graphX/edges$ ls
part-00000-48027420-a234-46c4-9b22-5f4f9a34c2a0-c000.snappy.parquet  _SUCCESS
hadoop@master:~/spark-graphX/edges$ ls -l
total 4
-rw-r--r-- 1 hadoop hadoop 802 May 12 03:42 part-00000-48027420-a234-46c4-9b22-5f4f9a34c2a0-c000.snappy.parquet
```

图 10-4　parquet 格式保存的运行结果

Json 格式的图保存，示例代码如下。

```
>>>graph.vertices.write.json("/home/hadoop/spark-graphX/vertices_json")
>>>ls
>>>part-00000-7a9d9acc-50a6-43ec-8b4b-3f29e8546a98-c000.json   _SUCCESS
```

Json 格式保存的运行结果如图 10-5 所示。

图 10-5　json 格式保存的运行结果

2. 加载图

图加载可以理解为对图的顶点和边的持久化，以便随时调用和加载。示例代码如下。

```
>>> v = spark.read.parquet("hdfs:// xdata-m1:8020/user/ua50/vertices")
>>> e = spark.read.parquet("hdfs:// xdata-m1:8020/user/ua50/edges")
>>> newGraph = GraphFrame(v, e)
```

10.3 编程实现——基于 GraphFrames 的网页排名

【任务描述】使用 PageRank 算法，基于 GraphFrames 实现网页排名。

PageRank 是 Google 公司用来标识网页的等级/重要性的一种算法，是 Google 公司用来衡量网站好坏的一种标准。Google 公司通过 PageRank 算法来调整结果，使等级更高、重要性更强的网页在搜索结果中越靠前，从而提高搜索结果的相关性和质量。

PageRank 算法是一种通过网页链接计算网页得分，进而对网页进行排名的算法，其基本思想为：如果一个网页被很多其他网页链接到，则说明这个网页比较重要，该网页的 PageRank 值也会相对较高；如果一个 PageRank 值很高的网页链接到一个其他网页，那么被链接的网页的 PageRank 值会相应地提高。

10.3.1　准备数据集

现在有两份数据：一份为网页信息数据 GraphX-wiki-vertices.txt，该数据可作为顶点数据，其中包括顶点编号和网页标题；另一份为网页之间的链接关系数据 GraphX-wiki-edges.txt，该数据可作为边数据，由两个顶点 ID 构成。部分数据如表 10-1 和表 10-2 所示。

表 10-1　GraphX-wiki-vertices.txt 部分数据

顶点编号	网页标题
6598434222544540151	Adelaide Hanscom Leeson
7814958205460279317	David Dodge (novelist)
3858831448322232257	Howard League for Penal Reform
1778261942684788432	ChelseaQuinnYarbro
4201849915685975228	DominickMontiglio
7447657998247897725	Joseph W. Estabrook
7816201114341688960	Spencer J. Palmer
290343697355110541	Aquatic Park (Berkeley)
1748906252821810168	JohanGaltung
3683742251178101007	Henry Howard, 7th Duke of Norfolk

表 10-2　GraphX-wiki-edges.txt 部分数据

顶点编号	关联关系顶点编号
36359329835505530	6843358505416683693
168437400931144903	961421098734626813
168437400931144903	1367968407401217879
168437400931144903	2270437664547777682

续表

顶点编号	关联关系顶点编号
168437400931144903	2381426201672413470
168437400931144903	3694025250309277803
168437400931144903	4289177408495534056
168437400931144903	4876378461651498574

10.3.2 GraphFrames 实现算法

1．广度优先搜索算法

官方接口为 bfs，代码如下。

```
bfs(fromExpr, toExpr, edgeFilter = None, maxPathLength = 10)
```

在上述代码中，fromExpr 参数表示 Spark SQL 表达式，指定搜索起点；toExpr 参数表示 Spark SQL 表达式，指定搜索终点；edgeFilter 参数指定搜索过程需要忽略的边，也是 Spark SQL 表达式；maxPathLength 参数表示路径的最大长度，若搜索结果路径长度超过该值，则算法终止。

广度优先搜索的示例代码如下。

```
paths = graph.bfs("name = 'Alice'", "age > 34")
paths.show()
```

2．最短路径算法

官方接口为 shortestPaths，代码如下。

```
shortestPaths(landmarks)
```

GraphFrames 中最短路径算法实际上是通过封装 GraphX 的最短路径算法来实现的，GraphX 实现的是单源最短路径算法，采用经典的 Dijkstra 算法。虽然算法命名是最短路径，但 GraphFrames 最短路径算法的返回结果只有距离值，并不会返回完整的路径。上述代码中的 landmarks 参数表示要计算的目标顶点 ID 集。最短路径算法的示例代码如下。

```
# landmarks is vector of target vertices
paths = graph.shortestPaths(landmarks = ["a", "d"])
paths.show()
```

最短路径算法可以计算图中的每个顶点到目标顶点的最短距离，但是会忽略边的权重。

3．三角形计数算法

三角形计数算法用于确定通过图数据集中每个顶点的三角形数量，其示例代码如下。当计算三角形个数时，图会被作为无向图进行处理，平行边仅计算一次，自环则会被忽略。三角形计数算法在社交网络分析中被大量使用。一般来说，在一个网络中，三角形个数越多，这个网络连接越紧密。例如，一个重要的统计特征——全局聚类系数就是基于三角形数量计算的，它是衡量社交网站中本地社区凝聚力的重要参考标准。

```
results = graph.triangleCount()
results.select("id", "count").show()
```

4．连通分量算法

连通分量算法可用于发现网络中的环，经常用于社交网络中来发现社交圈子。该算法使用顶点 ID 标注图中的每个连通体，将连通体中序号最小的顶点的 ID 作为连通体的 ID。连通分量算法的示例代码如下。

```
spark.sparkContext.setCheckpointDir('checkpoint')
results = graph.connectedComponents()
results.select("id", "component").orderBy("component").show()
results = graph.stronglyConnectedComponents(maxIter = 10)
results.select("id", "component").orderBy("component").show()
```

连通分量算法忽略边的方向，将图视为无向图。GraphFrames 还提供了强连通分量算法，它可以接收 maxIter 参数，该参数用于指定最大迭代次数。

5．标签传播算法

标签传播算法最早是针对社区发现的问题提出的一种解决方案。社区是一个模糊的概念，一般来说，社区是指一个子图，其内部顶点之间连接紧密，而与其他社区顶点之间连接稀疏。根据各社区顶点有无交集，又可分为非重叠型社区和重叠型社区。标签传播算法适用于非重叠社区，该算法的示例代码如下。

```
results = graph.labelPropagation(maxIter = 5)
results.show()
```

标签传播算法的优点是简单快捷、时间复杂度低、接近线性时间，缺点是结果不稳定。它的基本步骤如下。

步骤 1：初始时，给每个节点指定一个唯一的标签。

步骤 2：每个节点使用其邻居节点标签中最多的标签来更新自身的标签。

步骤 3：反复执行步骤 2，直到每个节点的标签都不再发生变化为止。

6．PageRank 算法

PageRank 算法最初被拉里·佩奇和谢尔盖·布林用来解决搜索引擎中网页排名的问题，故又称网页排名算法。该算法可以用于评估有向图中顶点的重要性，例如评估社交网络中关注度高的用户。与三角形计数算法相比，PageRank 算法是相关性的度量，而三角形计数算法是聚类的度量。PageRank 算法的示例代码如下。

```
results = graph.pageRank(resetProbability = 0.15, maxIter = 10)
results.vertices.show()
results = graph.pageRank(resetProbability = 0.15, tol = 0.01)
results.vertices.show()
results.edges.show()
```

PageRank 算法的计算过程如下。

① 初始化图中顶点的 PageRank 值为 $1/n$，其中的 n 表示图中顶点的个数。

② 按照以下步骤进行迭代。

步骤 1：每个顶点将其当前的 PageRank 值平均分配到顶点的出边上，即 PageRank 值$/m$，其中的 m 表示顶点的出度。

步骤 2：对每个顶点入边的 PageRank 值求和，得到顶点的新的 PageRank 值。

③ 如果相较于上一轮循环，整个图中顶点的 PageRank 值没有明显改变，即 PageRank

值趋于稳定，则算法结束。

```
pageRank(resetProbability = 0.15, sourceId = None, maxIter = None, tol = None)
```

在上述代码中，resetProbability 参数表示算法里的常数 alpha，默认值为 0.15；sourceId 表示顶点 ID，用于个性化 PageRank 算法，该参数可选；maxIter 参数表示迭代的最大次数；tol 参数表示最终收敛的公差值。

10.3.3 使用 PageRank 进行网页排名

使用 PageRank 算法进行网页排名的代码如下。

```
from pyspark.sql.types import *
from pyspark.sql import SparkSession

spark = SparkSession.builder.config(
'spark.jars.packages',
                'graphframes:graphframes:0.8.0-spark3.0-s_2.12').getOrCreate()
# filePath = 'datas/GraphX-wiki-vertices.txt'
filePath = 'datas/GraphX-wiki-edges.txt'
schema = StructType([StructField('src', LongType(), True), StructField(
        'dst', LongType(), True)])
edgesDF = spark.read.load(filePath, format = 'csv', schema = schema,
delimiter = '\t', mode = 'DROPMALFORMED')
edgesDF.cache()
srcDF = edgesDF.select(edgesDF.src).distinct()
distDF = edgesDF.select(edgesDF.dst).distinct()
verticesDF = srcDF.union(distDF).distinct().withColumnRenamed('src', 'id')
verticesDF.cache()
from graphframes import GraphFrame
graph = GraphFrame(verticesDF, edgesDF)
ranks = graph.pageRank(resetProbability = 0.15, maxIter = 5)
ranks.vertices.show()
ranks.edges.show()
```

10.4 本章小结

本章介绍了图的基本概念、图计算的应用和 GraphFrames，帮助读者了解 GraphFrames 编程模型。本章还展示了一个基于 GraphFrames 网页排名项目的编程实现过程，使读者对 GraphFramesX 的基本使用及应用 GraphFrames 解决实际问题有了更加深入的理解。

第 11 章 大数据生态常用工具介绍

数据采集是大数据分析全流程的重要环节，典型的数据采集工具包括 ETL 工具、日志采集工具、数据迁移工具等。本章将介绍具有代表性的数据采集工具，其中包括日志采集工具 Flume、Kafka 和数据迁移工具 Sqoop，详细介绍这些工具的安装和使用方法，同时通过相关实例来加深读者对工具作用及其使用方法的理解。

【学习目标】

1．掌握 Flume 安装及使用方法。

2．掌握 Kafka 安装及使用方法。

3．掌握 Sqoop 安装及使用方法。

4．通过编写 Spark 程序使用 Kafka 数据源。

11.1 Flume 的安装与使用

【任务描述】掌握 Flume 的安装和使用方法。

11.1.1 安装及配置 Flume

Flume 的安装过程如下。

1．下载安装文件

登录 Linux 操作系统，在 Linux 操作系统中（不是 Windows 操作系统）中打开浏览器，在 Apache 官网上下载 Flume 的安装文件。读者也可以从本书随书资源中获取 Flume 安装包 apache-flume-1.7.0-bin.tar.gz。浏览器会默认把下载文件都保存到当前用户的下载目录中，由于本书全部采用 hadoop 用户登录 Linux 操作系统，所以 apache-flume-1.7.0-bin.tar.gz 安装包会被保存到"/home/hadoop/下载/"目录下。

需要注意的是，如果是在 Windows 操作系统中下载安装包 apache-flume-1.7.0-bin.tar.gz，则需要通过 FTP 软件上传到 Linux 操作系统的"/home/hadoop/下载/"目录下，这个目录是本书所有安装包的中转站。

安装包下载完后，需要进行解压。按照 Linux 操作系统使用的默认规范，用户安装的软件一般存储在/usr/local/目录下。使用 Hadoop 用户登录 Linux 操作系统，打开一个终端，执行以下命令。

```
$cd~
$sudo tar -zxvf ./下载/apache-flume-1.7.0-bin.tar.gz -C /usr/local
```

注意：将 apache-flume-1.7.0-bin.tar.gz 解压到/usr/local 目录下的命令中一定要加上-C，否则会出现错误。

然后将解压后文件的名称修改为 flume，这样就不用每次都要输入很长的文件名，命令如下。

```
$cd /usr/ local
$sudo mv ./apache-flume- 1.7.0-bin ./flume
```

2．配置环境变量

首先使用 vim 编辑器打开~/.bashrc 文件，命令如下。

```
$sudo vim~/.bashrc
```

然后在该文件开头加入以下代码。

```
export JAVA_HOME = /usr/lib/jvm/java-8-openjdk-amd64;
export FLUME_HOME = /usr/local/flume
export FLUME_CONF_DIR = $FLUME_HOME/confexport
PATH = $PATH:SFLUME_HOME/bin
```

需要注意的是，本书中的各个章节都对~/.bashrc 文件进行了修改。如果之前该文件中已经增加了其他内容，那么这里不要删除这些内容，把上面 4 行内容继续增加到该文件开头即可。另外，如果之前该文件中已经加入“export JAVA_HOME=/usr/lib/jvm/java-8-openjdk- amd64;”这行语句，那么，这里就不用重复添加这行语句。

接下来执行以下命令使环境变量立即生效。

```
$source ~/.bashrc
```

最后修改配置文件 flume-env.sh，命令如下。

```
$cd /usr/local/flume
$sudo mvflume -env.sh.template flume-env.sh
$sudo vim flume-env.sh
```

在 flume-env.sh 文件开头加入以下语句。

```
export JAVA HOME = /usr/lib/jvm/java-8-openjdk-amd64;
```

3．启动 Flume

执行以下语句启动 Flume。

```
$cd /usr/local/flume
$./bin/flume- ng version
```

启动成功后会出现以下信息。

```
flume 1.7,0
Source code repository: https://×××.org/repos/asf/fluwe.gitRevision:
511d868555dd4d16eoce4fedc72c2d1454546707
Conpiled by bessbd on wed oct 12 20:51;10 CEST 2016
From source with checksun 0d21b3ffdc55a07e1d08875872c00523
```

需要注意的是，如果系统中之前已经安装 HBase，那么按照上述方法配置并启动 Flume 后会出现“找不到或无法加载主类 org.apache.flume.tools.GetJavaProper”的错误。此时，我们可以通过修改 hbase-env.sh 文件来解决这个错误。首先使用 vim 编辑器打开 hbase-env.sh 文件，命令如下。

```
$cd /usr/local/hbase/conf
$sudo vim hbase-env.sh
```

在 hbase-env.sh 文件中找到下面展示的这一行内容，把这行内容注释掉，即在语句前面加上一个#。

```
export HBASE_CLASSPATH = /home/hadoop/hbase/conf
```

添加后的效果如下。

```
# export HBASE_CLASSPATH = /home/hadoop/hbase/conf
```

之后保存该文件并退出 vim 编辑器。通过这种方式，我们就可以顺利启动 Flume。

11.1.2　实例分析

1．使用 Flume 接收来自 AvroSource 的信息

AvroSource 可以发送一个给定的文件给 Flume，Flume 接收到该文件后可以进行相应的处理，例如显示到屏幕上。

① 创建 Agent 配置文件。首先使用以下命令在/usr/local/flume/conf 目录下创建一个新文件 avro.conf。

```
$cd /usr/local/flume
$sudo vim ./conf/avro.conf
```

然后在 avro.conf 文件中写入以下内容。

```
a1.sources = r1
a1.sinks = k1
a1.channels = c1
# 配置 Source
a1.sources.r1.type = avro
a1.sources.r1.bind = 0.0.0.0
a1.sources.r1.port = 4141
# 配置 Sink
a1.sinks.k1.type = logger
# 配置 Channel
a1.channels.c1.type = memory
a1.channels.c1.capacity = 1000
a1.channels.c1.transactionCapacity = 100
# 绑定 Source 和 Sink 到 Channel
a1.sources.r1.channels = c1
a1.sinks.k1.channel = c1
```

② 启动 Flume Agent a1。执行以下命令启动日志控制台。

```
$/usr/local/flume/bin/flume-ng agent --conf /home/hadoop/flume/conf/
--conf-file /home/hadoop/flume/ conf/avro.conf  --name a1 -Dflume.root.logger =
INFO,console
```

③ 使用以下命令打开新终端，向文件 log.00 输入一些信息。

```
echo "hello world"> /home/hadoop/flume/log.00
```

④ 使用 avro-client 发送文件，具体如下。

```
flume-ng avro-client -c /home/hadoop/flume/conf/ -H 0.0.0.0 -p 4141 -F /home/
hadoop/flume/log.00
```

⑤ 在输出信息的最后一行可看到“hello world”，如图 11-1 所示。

```
18/08/22 12:50:30 INFO sink.LoggerSink: Event: { headers:{} body: 68 65 6C 6C 6F
 20 77 6F 72 6C 64                          hello world }
```

图 11-1　输出信息

2. 使用 Syslogtcp Source 接收外部数据源

使用 Syslogtcp Source 接收外部数据源，HDFS 作为 Sink，将数据缓存在 Memory Channel，保存到 HDFS 中。

① 创建 Agent 配置文件/home/hadoop/flume/conf/syslogtcp.conf，并按以下内容进行配置。

```
a1.sources = r1
a1.sinks = k1
a1.channels = c1
# 配置 Source
a1.sources.r1.type = syslogtcp
a1.sources.r1.port = 5140
a1.sources.r1.host = localhost
# 配置 Sink
a1.sinks.k1.type = hdfs
a1.sinks.k1.hdfs.path = hdfs:// 192.168.30.128:8020/user/hadoop/flume/syslogtcp
a1.sinks.k1.hdfs.filePrefix = Syslog
a1.sinks.k1.hdfs.round = true
a1.sinks.k1.hdfs.roundValue = 10
a1.sinks.k1.hdfs.roundUnit = minute
# 配置 Channel
a1.channels.c1.type = memory
a1.channels.c1.capacity = 1000
a1.channels.c1.transactionCapacity = 100
# 绑定 Source 和 Sink 到 Channel
a1.sources.r1.channels = c1
a1.sinks.k1.channel = c1
```

② 启动 Agent a1，命令如下。

```
Flume-ng agent -c /home/hadoop/flume/conf/-f  /home/hadoop/flume/conf/syslogtcp.conf
-n a1 -Dflume.root.logger = INFO,console
```

③ 测试是否产生 syslogtcp，命令如下。

```
echo "hello syslogtcp" | nc localhost 5140
```

④ 查看 HDFS 上/user/hadoop/flume/是否生成了 syslogtcp 文件，并查看内容。如果生成了该文件，则在输出信息中能看到“hello syslogtcp”。

11.2 Kafka 的安装与使用

【任务描述】掌握 Kafka 的安装与使用方法。

Kafka 是一种高吞吐量的分布式发布订阅消息系统，它可以处理消费者规模的网站中的所有动作流数据。Kafka 的目的是通过 Hadoop 和 Spark 的并行加载机制来统一线上和离线消息的处理。

11.2.1　Kafka 相关概念

为了更好地理解和使用 Kafka，我们先介绍 Kafka 的相关概念。

① Broker：Kafka 集群包含一个或多个服务器，这些服务器被称为 Broker。

② Topic：每条发布到 Kafka 集群的消息都有一个类别，这个类别被称为 Topic。物理上不同 Topic 的消息分开存储，逻辑上一个 Topic 的消息虽然保存在一个或多个 Broker 上，但用户只需要指定消息的 Topic，即可生产或消费数据，而不必关心数据存储在何处。

③ Partition：是物理上的概念，每个 Topic 包含一个或多个 Partition。

④ Producer：负责发布消息到 Kafka Broker 。

⑤ Consumer：消息消费者，向 Kafka Broker 读取消息的客户端。

⑥ Consumer Group：每个 Consumer 属于一个特定的 Consumer Group，可为每个 Consumer 指定 groupname，若不指定 groupname，则属于默认的 group。

11.2.2　安装 Kafka

在 Kafka 官网下载 Kafka 稳定版本的安装包 kafka_2.12-0.10.2.2.tgz，此安装包内已经附带 Zookeeper，不需要额外安装 Zookeeper。读者也可以从本书配套资源中获取 Kafka 的安装包 kafka_2.12-0.10.2.2.tgz。

安装包下载完后需要进行解压。按照 Linux 操作系统使用的默认规范，用户安装的软件一般存储在/usr/local/目录下。使用 Hadoop 用户登录 Linux 操作系统，打开一个终端，执行以下命令。

```
$cd～/下载
$sudo tar -zxf kafka_2.12-0.10.2.2.tgz -C/usr/local
$cd /usr/local
$sudo mv kafka_2.12-0.10.2.2/ ./kafka
$sudo chown-R hadoop ./ kafka
```

11.2.3　实例分析

新建一个 Linux 操作系统终端，执行以下命令来启动 Zookeeper。

```
$cd /usr/local/kafka
$bin/zookeeper-server-start.sh config/zookeeper.properties
```

执行完上面命令后，终端窗口会返回一堆信息，之后就停住不动，没有回到 Shell 命令提示符状态，注意这不是死机了，而是 Zookeeper 服务器启动了，正处于服务状态。因此，千万不要关闭这个终端窗口，一旦关闭，Zookeeper 服务就会停止。

新建第二个终端，输入以下命令来启动 Kafka。

```
$cd /usr/ local/kafka
$ ./bin/kafka-server-start.sh config/server.properties
```

同样地，执行完上面命令后，终端窗口会返回一堆信息，然后就停住不动了，没有回到 Shell 命令提示符状态，注意这不是死机了，而是 Kafka 服务器启动了，正处于服务状态。因此，千万不要关闭这个终端窗口，一旦关闭，Kafka 服务就会停止。

新建第三个终端，输入以下命令。

```
$cd /usr/local/kafka
$./bin/kafka-topics.sh-- create- -zookeeper localhost:2181--replication-factor
1--partitions 1--topic dblab
```

上述命令以单节点的配置方式创建了一个名为 dblab 的 topic。我们可以用 list 命令列出所有创建的 topics，来查看刚才创建的 topic 是否存在，命令如下。

```
$cd /usr/local/kafka
$ ./bin/kafka-topics.sh--list--zookeeper localhost:2181
```

从结果中可以看到，名为 dblab 的 topic 已经存在。接下来我们用 producer 生产一些数据，命令如下。

```
$cd /usr/local/kafka
$./bin/kafka-console-producer.sh--broker-list localhost:9092--topic dblab
```

上述命令执行后，我们可以在该终端上输入以下信息作为测试。

```
hello hadoop
hello xmu
hadoop world
```

之后新建第四个终端，输入以下命令，以使用 consumer 来接收数据。

```
$cd /usr/local/kafka
$./bin/kafka-console-consumer.sh--zookeeper localhost:2181--topic dblab--from-
beginning
```

执行该命令以后，我们就可以看到刚才在另一个终端的 producer 产生的 3 条信息“hello hadoop”“hello xmu”和“hello world”，这说明 Kafka 安装成功了。

11.3 Sqoop 的安装与使用

【任务描述】掌握 Sqoop 的安装与使用方法。

Sqoop 是一款开源工具，主要用于在 Hadoop 与传统的关系数据库之间进行数据的传递。Sqoop 可以将一个关系数据库（例如 MySQL、Oracle、PostgreSQL 等）中的数据导入 Hadoop 的 HDFS、HBase、Hive 中，也可以将 Hadoop 的数据导入关系数据库中。

Sqoop 项目开始于 2009 年，最早是作为 Hadoop 的一个第三方模块存在，后来为了让使用者能够快速部署，也为了让开发人员能够更快地迭代开发，Sqoop 独立成为一个 Apache 项目。

Sqoop 的安装环境如下。

① 操作系统：Linux 操作系统（建议 Ubuntu）。

② Sqoop：1.4.7 版本。

③ Hadoop：3.2.0 版本。

④ MySQL：5.7.15 版本。

需要注意的是，Sqoop 官方停止了更新维护，支持的 Hadoop 版本停留在了 hadoop 2.6，所以我们需要通过以下方法来兼容 Hadoop 3.X 版本。

11.3.1 安装及配置 Sqoop

1．下载安装文件

读者先访问 Sqoop 官网，在该网站上下载 Sqoop 安装包 sqoop-1.4.7. bin__hadoop-2.6.0.tar.gz 和 sqoop-1.4.7.tar.gz。

将 sqoop-1.4.7.tar.gz 上传到虚拟机，之后对该文件进行解压。执行以下命令，完成安装包的解压缩。

```
$tar -zxvf sqoop-1.4.7.tar.gz -C /usr/local
```

提取 sqoop-1.4.7.bin__hadoop-2.6.0.tar.gz 安装包中的 sqoop-1.4.7.jar，将其复制并粘贴到 sqoop-1.4.7 根目录，如图 11-2 所示。纯净版 Sqoop 是没有这个 jar 包的。

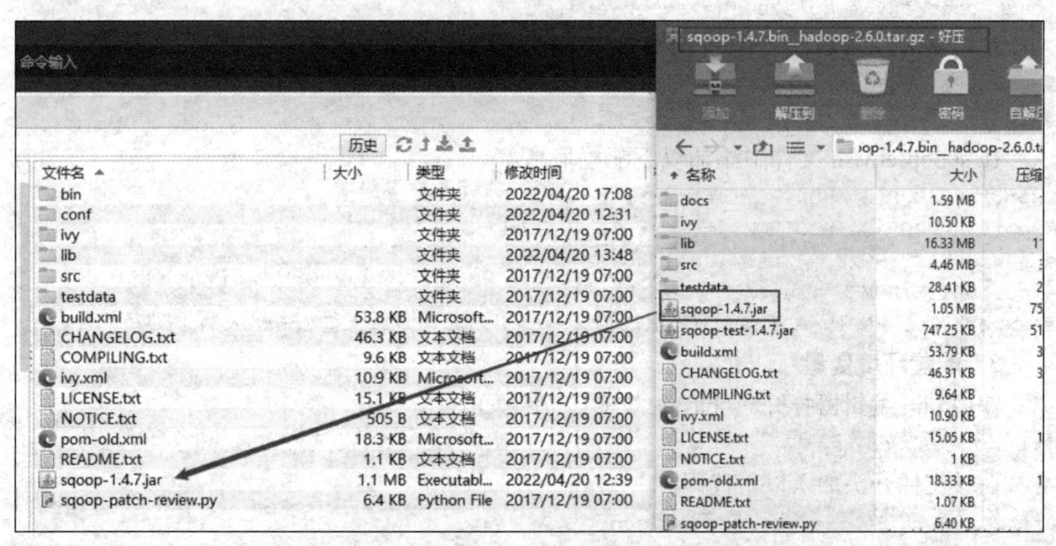

图 11-2 复制并粘贴 sqoop-1.4.7.jar 到 sqoop-1.4.7 根目录

将 lib 目录的 3 个 jar 文件复制并粘贴到 sqoop-1.4.7/lib/目录下，如图 11-3 所示。正常纯净版 sqoop 的 lib 目录下是没有文件的。

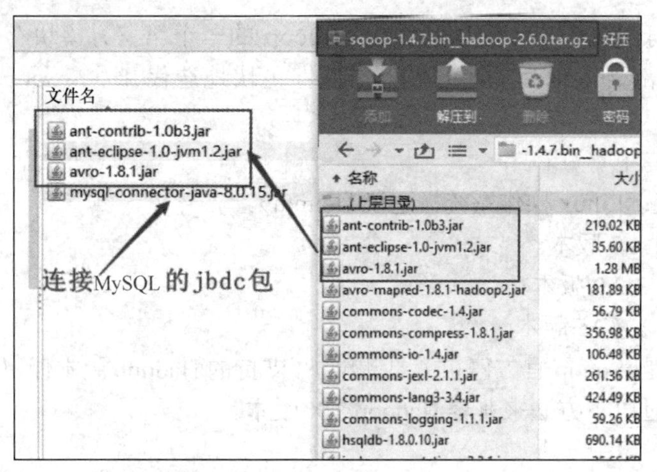

图 11-3　复制其他 jar 放到 sqoop-1.4.7 的 lib 目录

将已解压的文件名修改为 sqoop，这样就不用每次都输入很长的文件名，命令如下。

```
$cd /usr/local
$sudo mv sqoop-1.4.7 sqoop
```

接下来把 sqoop 文件夹的权限赋予当前的 Hadoop 用户，命令如下。

```
$sudo chown-R hadoop:hadoop sqoop
```

2. 修改配置文件

执行以下命令，复制/usr/local/sqoop/conf 目录下的配置文件 sqoop-env-template.sh，并将复制文件命名为 sqoop-env.sh。

```
$cd sqoop/conf/
$cat sqoop-env-template.sh>>sqoop-env.sh
```

使用 vim 编辑器打开 sqoop-env.sh 文件进行编辑，命令如下。

```
$cd /usr/local/sqoop/conf/
$vim sqoop-env.sh
```

在 sqoop-env.sh 文件中添加以下配置信息。

```
Export HADOOP_COMMON_HOME = /usr/local/hadoop
export HADOOP_MAPRED_HOME = /usr/local/hadoop
export HBASE_HOME = /usr/local/hbase
export HIVE_HOME = /usr/local/hive
# export ZOOCFGDIR = # 如果读者配置了 Zookeeper，这里也要配置 Zookeeper 的路径
```

3. 配置环境变量

使用 vim 编辑器打开～/.bashrc 文件，命令如下。

```
$vim~/.bashrc
```

在该文件开头加入以下代码。

```
export SQOOP_HOME = /usr/local/sqoop
export PATH = $PATH:$SBT_HOME/bin:$SQOOP_HOME/bin
export CLASSPATH = $CLASSPATH:$SQOOP_HOME/lib
```

保存该文件，并退出 vim 编辑器。接下来执行以下命令，使环境变量立即生效。

```
$source~/.bashrc
```

11.3.2　添加 MySQL 驱动程序

Sqoop 经常与 MySQL 结合，从 Hadoop 数据源向 MySQL 数据库导入数据，或者从 Hadoop 内的各个组件导出数据到 MySQL，所以我们要为 Sqoop 配置 Java 连接器（MySQL 的 Java 连接器也称为 JDBC 驱动）。用户可以从 MySQL 官网下载对应的连接器，这里下载的是 mysql-connector-java-5.1.46.tar.gz，并将该文件存储在用户主目录下。运行以下命令来解压这个文件。

```
$tar -xzvf mysql-connector-java-5.1.46.tar.gz
$ls # 这时就可以看到解压缩后的目录mysql-connector-java-5.1.46
$cp ./mysql-connector-java-5.1.46/mysql-connector-java-5.1.46-bin.jar
/usr/local/sqoop/lib
```

MySQL 连接器文件夹及文件如图 11-4 所示。

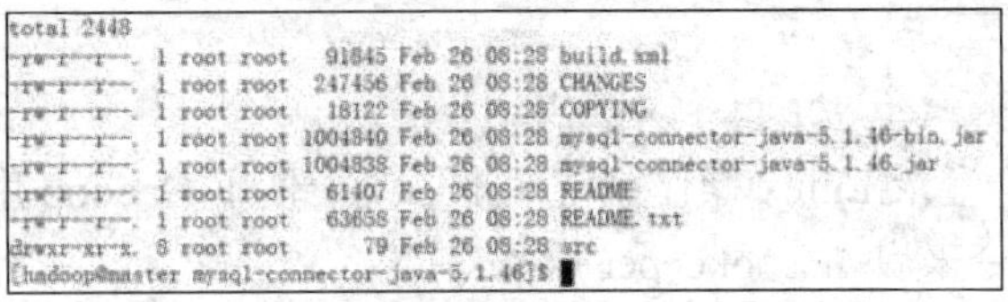

```
total 2448
-rw-r--r--. 1 root root   91845 Feb 26 08:28 build.xml
-rw-r--r--. 1 root root  247456 Feb 26 08:28 CHANGES
-rw-r--r--. 1 root root   18122 Feb 26 08:28 COPYING
-rw-r--r--. 1 root root 1004840 Feb 26 08:28 mysql-connector-java-5.1.46-bin.jar
-rw-r--r--. 1 root root 1004838 Feb 26 08:28 mysql-connector-java-5.1.46.jar
-rw-r--r--. 1 root root   61407 Feb 26 08:28 README
-rw-r--r--. 1 root root   63658 Feb 26 08:28 README.txt
drwxr-xr-x. 8 root root      79 Feb 26 08:28 src
[hadoop@master mysql-connector-java-5.1.46]$
```

图 11-4　MySQL 连接器文件夹及文件

11.3.3　测试 Sqoop 与 MySQL 的连接

首先，要确保 MySQL 服务已经启动。如果 MySQL 服务没有启动，则执行以下命令进行启动。

```
$service mysql start
```

然后，测试 Sqoop 与 MySQL 之间的连接是否成功，命令如下。

```
$sqoop list- databases - - connect jdbc:mysql :// 127.0.0.1:3306/
--usernameroot -P
```

执行以下命令，在该命令的运行结果中可以看到图 11-5 所示内容，这表示安装成功。

```
$sqoop help
```

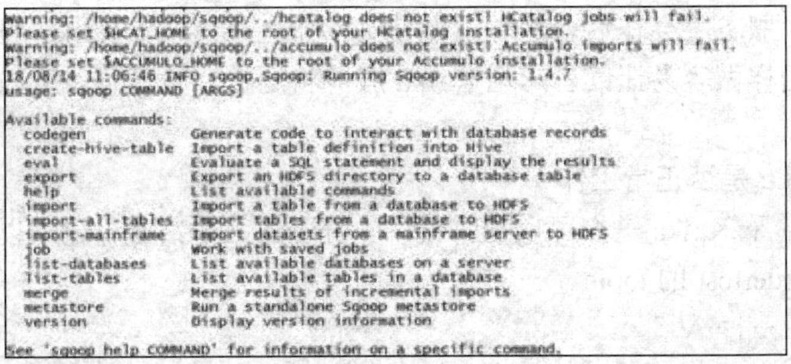

```
Warning: /home/hadoop/sqoop/../hcatalog does not exist! HCatalog jobs will fail.
Please set $HCAT_HOME to the root of your HCatalog installation.
Warning: /home/hadoop/sqoop/../accumulo does not exist! Accumulo imports will fail.
Please set $ACCUMULO_HOME to the root of your Accumulo installation.
18/08/14 11:06:46 INFO sqoop.Sqoop: Running Sqoop version: 1.4.7
usage: sqoop COMMAND [ARGS]

Available commands:
  codegen            Generate code to interact with database records
  create-hive-table  Import a table definition into Hive
  eval               Evaluate a SQL statement and display the results
  export             Export an HDFS directory to a database table
  help               List available commands
  import             Import a table from a database to HDFS
  import-all-tables  Import tables from a database to HDFS
  import-mainframe   Import datasets from a mainframe server to HDFS
  job                Work with saved jobs
  list-databases     List available databases on a server
  list-tables        List available tables in a database
  merge              Merge results of incremental imports
  metastore          Run a standalone Sqoop metastore
  version            Display version information

See 'sqoop help COMMAND' for information on a specific command.
```

图 11-5　Sqoop 启动信息

11.4 编程实现——编写 Spark 程序使用 Kafka 数据源

【任务描述】Spark Streaming 是用于进行流计算的组件。Kafka 可以作为数据源，让它产生数据并发送给 Spark Streaming 应用程序，Spark Streaming 应用程序再对接收到的数据进行实时处理，从而完成一个典型的流计算过程。

为了让 Spark Streaming 应用程序能够顺利使用 Kafka 数据源，在安装 Kafka 的时候一定要注意，在 Kafka 官网上下载安装包时，一定要选择和自己计算机上已经安装的 Scala 版本号一致的安装包。本书安装的 Spark 版本号是 3.3.0，Scala 版本号是 2.12，所以，读者一定要选择 Kafka 版本号是 2.12 开头的安装包。

11.4.1 Kafka 准备工作

1. 启动 Kafka

首先启动 Kafka。登录 Linux 操作系统（本书统一使用 Hadoop 用户登录），打开第一个终端，输入以下命令来启动 Zookeeper 服务。

```
$cd /usr/local/kafka
$./bin/zookeeper-server-start.sh config/zookeeper.properties
```

执行上面的命令以后，终端窗口会返回一堆信息，然后就停住不动了，没有回到 Shell 命令提示符状态，注意这不是死机了，而是 Zookeeper 服务启动了，正处于服务状态。千万不要关闭这个终端窗口，一旦关闭，Zookeeper 服务就会停止。

打开第二个终端，输入以下命令来启动 Kafka 服务。

```
$cd /usr/local/kafka
$./bin/kafka-server-start.sh config/server.properties
```

同样地，执行完上面命令后，终端窗口会返回一堆信息，然后就停住不动了，没有回到 Shell 命令提示符状态，注意这不是死机了，而是 Kafka 服务启动了，正处于服务状态。千万不要关闭这个终端窗口，一旦关闭，Kafka 服务就会停止。

当然，还有一种方式是采用加了&的命令，具体如下。

```
$cd /usr/local/kafka
$./bin/kafka-server-start.sh config/server.properties &
```

这样，Kafka 服务将会在后台运行。即使关闭了上述终端，Kafka 服务也会一直在后台运行。不过，这样做容易让读者忘记 Kafka 服务将还在后台运行，所以我们建议暂时不要用&。

2. 测试 Kafka 是否正常工作

下面先测试一下 Kafka 是否可以正常使用。打开第三个终端，输入以下命令，创建一个名称为 wordsendertest 的 topic。

```
$cd /usr/local/kafka
$./bin/kafka-topics.sh- - create- - zookeeper localhost:2181--replication-factor
1--partitions 1--topic wordsendertest
```

```
# 这个 topic 名为 wordsendertest；2181 是 Zookeeper 默认的端口号； partition 是 topic
# 中的分区数；replication-factor 是备份的数量，常用在 Kafka 集群中，这里单机版就不用备份了
# 我们可以用 list 命令列出所有创建的 topics，查看上面创建的 topic 是否存在
$./bin/kafka-topics.sh--list--zookeeper localhost:2181
```

这个名为 wordsendertest 的 topic 是专门负责采集并发送一些单词的。下面我们用 producer 产生一些数据，在当前终端内继续输入以下命令。

```
$./bin/kafka - console - producer. sh - - broker - list localhost: 9092- -topic
 wordsendertest
```

执行完上面命令后，我们可以在当前终端内用键盘输入一些英文单词，例如输入以下内容。

```
hello hadoop
hello spark
```

这些单词是数据源，会被 Kafka 捕捉到并发送给消费者。现在可以启动一个消费者，查看刚才用 producer 产生的数据。打开第四个终端，输入以下命令。

```
$cd /usr/local/kafka
$./bin/kafka - console - consumer. sh - - zookeeper localhost: 2181 - -
topicwordsender--from-beginning
```

可以看到，屏幕上会显示出以下结果，这也是刚才在另一个终端中输入的内容。

```
hello hadoop
hello spark
```

到这里，与 Kafka 相关的准备工作就顺利结束了。

注意：所有终端都不要关闭，要继续留着后面使用。

11.4.2　Spark 准备工作

1．添加相关 jar 包

Kafka 和 Flume 等高级输入源需要依赖独立的库（jar 文件）。按照前文中已经安装好的 Spark 版本，这些所依赖的库都不在里面。为了证明这一点，我们先进行测试。打开一个新的终端，启动 Spark Shell，命令如下。

```
$cd /usr/local/spark
$./bin/spark-shell
```

启动成功后，在 Spark Shell 中执行以下语句。

```
scala> import org.apache.spark.streaming.kafka._
<console>:25: error: object kafka is not a member of package
org.apache.spark.streaming
import org.apache.spark.streaming.kafka._
```

可以看到，程序会报错，这是因为找不到相关的依赖库。我们需要下载 spark-streaming-kafka-0-10_2.12.jar。在 Linux 操作系统中打开浏览器，在 Spark 官网上下载 spark-streaming-kafka-0-10_2.12.jar 文件，其中，2.12 表示 Scala 的版本。下载页面会列出 spark-streaming-kafka0-10_2.12.jar 的很多版本，这里选择 3.3.0 版本（因为本书安装的 Spark 版本是 3.3.0）下载即可。下载后的文件会被默认保存在当前 Linux 操作系统登录用户的下载目录下。本书统一使用 Hadoop 用户名登录 Linux 操作系统，所以，文件下载后

会被保存到“/home/hadoop/下载”目录下。读者也可以访问本书的配套资源，获取spark-streaming-kafka-0-10_2.12.jar文件。现在需要重新打开一个终端，输入下命令，把这个文件复制到Spark目录的lib目录下。

```
$cd /usr/local/spark/lib
$mkdir kafka
$cd ~
$cd 下载
$cp ./spark-streaming-kafka-0-10_2.12.jar /usr/local/spark/lib/kafka
```

我们还需要在Spark官网下载文件spark-streaming_2.12-3.3.0.jar。读者也可以访问本书随书资源获取该文件，其中，2.12表示Scala版本号，3.3.0表示Spark版本号。

spark-streaming_2.12-3.3.0.jar下载成功以后，默认会被存储到当前Linux登录用户的下载目录下。本书统一使用Hadoop用户名登录Linux操作系统，所以，文件下载后会保存到“/home/hadoop/下载”目录下。重新打开一个终端，输入以下命令，把这个文件复制到Spark目录的lib目录下。

```
$cd /usr/local/spark/lib
$cd ~
$cd 下载
$cp ./spark-streaming 2.12-3.3.0.jar /usr/local/spark/lib/kafka
```

继续在终端中执行以下命令，把Kafka安装目录的lib目录下的所有jar文件复制到/usr/local/spark/lib/kafka目录下。

```
$cd /usr/local/kafka/libs
$ls
$cp ./* /usr/local/spark/lib/kafka
```

这时，有些jar文件会和Spark中已经有的jar文件发生冲突。程序一旦运行就会出现一堆错误，因此，我们需要把这些会发生冲突的jar文件删掉，命令如下。

```
$cd /usr/local/kafka/libs
$ls
$rm log4j*
$rm jackson*
```

2. 修改配置文件

现在我们配置/usr/local/spark/cont目录下的spark-env.sh文件，让Spark能够在启动的时候找到spark-streaming-kafka_2.12-3.3.0.jar等5个jar文件。使用vim编辑器打开spark-env.sh文件，命令如下。

```
$cd /usr/local/spark/conf
$vim spark-env.sh
```

因为这个文件之前已经反复修改过，目前该文件前面几行内容应该如下。

```
export SPARK_CLASSPATH = $SPARK CLASSPATH :/usr/local/spark/lib/hbase/*
export SPARK_DIST_CLASSPATH = $(/usr/local/hadoop/bin/hadoop classpath)
```

简单修改上述内容，把/usr/local/spark/lib/kafka/ *加进去。修改后的内容如下。

```
export SPARK_CLASSPATH = $SPARK CLASSPATH:/usr/local/spark/lib/hbase/*:/usr/
local/spark/lib/kafka/*
export SPARK_DIST_CLASSPATH = $(/usr/local/hadoop/bin/hadoop classpath)
```

保存该文件后，退出 vim 编辑器。

3. 启动 Spark Shell

执行以下命令启动 Spark Shell。

```
$cd /usr/local/spark
$./bin/Spark Shell
```

启动成功后，再次执行以下命令。

```
scala>import org.apache.spark.streaming.kafka._
// 会显示下面的信息
import org.apache.spark.streaming.kafka._
```

现在使用 import 语句时不会和之前一样出现错误信息了，这说明已经导入成功。至此，Spark 环境已经准备好，它可以支持 Kafka 相关编程。

11.4.3 编写代码

1. 编写生产者（producer）代码

打开一个终端，执行以下命令，创建代码目录和代码文件。

```
$cd /usr/local/spark/mycode
$mkdir kafka
$cd kafka
$mkdir -p src/main/scala
$cd src/main/scala
$vim KafkaWordProducer.scala
```

这里使用 vim 编辑器新建 KafkaWordProducer.scala。这是一个用于产生一系列字符串的程序，会产生随机的整数序列，每个整数被当成一个单词，提供给 KafkaWordCount 程序来进行词频统计。在 KafkaWordProducer.scala 中输入以下代码。

```
import java.util.HashMap
import org.apache. kafka. clients. producer.{ ProducerConfig,KafkaProducer,
ProducerRecord}
import org.apache.spark.streaming._
import org.apache.spark.streaming.kafka._
import org.apache.spark.Sparkconf
object KafkaWordProducer {
    def main(args : Array [String]){
        if(args.length<4){
            System.err.println (
            "Usage: KafkawordCountProducer<metadataBrokerList><topic>"+
            "<messagesPerSec><wordsPerMessage>")
            System.exit (1)
        }
        val Array (brokers,topic,messagesPerSec,wordsPerMessage) = args
        // zookeeper connection properties
        val props = new HashMap [String, Object]()
        props.put (ProducerConfig.BOOTSTRAP_SERVERS_CONFIG,brokers)
        props.put(ProducerConfig. VALUE_SERIAL工ZER_CLASS_CONFIG,
        "org.apache.kafka.common.serialization.stringSerializer")
```

```
        props.put (ProducerConfig.KEY_SERIALIZER_CLAsS_CONFIG,
        "org.apache.kafka.common.serialization.stringSerializer")
        val producer = new KafkaProducer [String,String](props)
    }
}
```

保存后退出 vim 编辑器。

2. 编写消费者（consumer）代码

继续在当前目录下创建 KafkaWordCount.scala 程序，命令如下。

```
$vim KafkaWordCount.scala
```

KafkaWordCount.scala 用于单词词频统计，会对 KafkaWordProducer.scala 发送的单词进行词频统计，代码如下。

```
import org. apache.spark._
import org.apache.spark.SparkConf
import org.apache.spark.streaming._
import org.apache.spark.streaming.kafka._
import org.apache.spark.streaming.streamingContext._
import org.apache.spark.streaming.kafka.Kafkattils
object KafkaWordCount{
    def main(args : Array [String]){
        StreamingExamples.setStreamingLogLevels ()
        val sc =
        new SparkConf().setAppName ( "KafkaWordCount").setMaster ("local[2]")
        val ssc = new StreamingContext(sc,Seconds(10))
        ssc.checkpoint("file:///usr/local/spark/mycode/kafka/checkpoint
        // 设置检查点，如果存储在 HDFS 上面，则写成类似 ssc.checkpoint (
        //"/user/Hadoop/checkpoint")这种形式，但是，要启动 hadoop
        val zkQuorum = "localhost:2181"// Zookeeper 服务器地址
        val group = "1"
        // topic 所在的 group，可以设置为自己想要的名称，例如不用 1，而是
        // val group = "test- consumer-group"
        val topics = "wordsender"// topics 的名称
        val numThreads = 1// 每个 topic 的分区数
        val topicMap = topics.split(", " ).map((_, numThreads.toInt) ).toMap
        val lineMap = KafkaUtils.createstream (ssc,zkQuorum, group, topicMap)
        val lines = lineMap.map(_._2)
        val words = lines.flatMap(_.split(""))
        val pair = words.map(x = >(x,1))
        val wordCounts = pair.reduceByKeyAndWindow(_+,- _,Minutes (2),
        Seconds (10),2)
        // 这行代码的含义在下一节的窗口转换操作中会有介绍
        wordCounts.print
        ssc.start
        ssc.awaitTermination
    }
}
```

保存后退出 vim 编辑器。

3. 编写日志格式设置代码

继续在当前目录下创建 StreamingExamples.scala 程序，命令如下。

```
$vim StreamingExamples.scala
```

下面是 StreamingExamples.scala 的代码，这段代码的功能是设置 log4j 的日志格式。

```
import org.apache.spark.Logging
import org.apache.log4j.{Level,Logger}
/**Spark Streaming 示例的实用函数*/
object streamingExamples extends Logging{
/** 设置合理的登录级别.*/
    def setstreamingLogLevels (){
        val log4jInitialized =
        Logger.getRootLogger.getAllAppenders.hasMore Elements
        if(! log4j Initialized){
            // 先初始化 Spark 的默认日志记录，再重写日志级别
        logInfo ("Setting log level to [WARN] for streaming example."+"
                To override add a custom log4j.properties to the classpath.")
            Logger.getRootLogger.setLevel (Level.WARN)
        }
    }
}
```

保存后退出 vim 编辑器。

4. 编译打包程序

至此，/usr/local/spark/mycode/kafka/src/main/scala 目录下有以下 3 个代码文件。

```
KafkaWordProducer.scala
KafkaWordCount.scala
StreamingExamples.scala
```

执行以下命令。

```
$cd /usr/local/spark/mycode/ kafka/
$vim simple.sbt
```

在 simple.sbt 中输入以下代码。

```
name : = "Simple Project"version : = "1, 0"
scalaversion : = "2.12"
libraryDependencies+ = "org.apache.spark" %"spark-core"%"3.3.0"
libraryDependencies+ = "org.apache.spark" %"spark-streaming_2.12" $"3.3.0"
libraryDependencies+ = "org.apache.spark" %"spark-streaming-kafka_2.12"%"3.3.0"
```

保存后退出 vim 编辑器。

执行以下命令，对上述代码进行打包和编译。

```
$cd /usr/local/spark/mycode/ kafka/
$/usr/local /sbt/sbt package
```

5. 运行程序

首先启动 Hadoop。如果前面 KafkaWordCount.scala 程序中采用了 ssc.checkpoint(" /user/hadoop/ checkpoint")形式，则表示检查点被写入了 HDFS，因此需要先启动 Hadoop。启动 Hadoop 的命令如下。

```
$cd /usr/local/hadoop
```

```
$./sbin/start-dfs.sh
```

之后就可以测试刚才生成的词频统计程序了。

要注意，之前已经启动了 Zookeeper 服务、Kafka 服务，因为之前那些终端窗口都没有关闭，所以，这些服务一直都在运行。如果不小心关闭了之前的终端窗口，读者需要参照前面的内容，再次启动 Zookeeper 服务、Kafka 服务。

打开一个终端，执行以下命令，运行 KafkaWordProducer 程序，生成一些单词（是一堆整数形式的单词）。

```
$cd /usr/local/spark
$/usr/local/spark/bin/spark-submit--class "KafkaWordProducer"
$/usr/local/spark/mycode/kafka/target/scala-2.12/simple-project2.12-1.0.jar
localhost:9092 wordsender 3 5
```

注意：在上面命令中，localhost : 9092 wordsender 3 5 是提供给 KafkaWordProducer 程序的 4 个输入参数。第 1 个参数 localhost:9092 表示 Kafka 的 broker 的地址。第 2 个参数 wordsender 表示 topic 的名称。在 KafkaWordCount.scala 代码中已经把 topic 名称固定为 wordsender，所以，KafkaWordCount 程序只能接收名称为 wordsender 的 topic。第 3 个参数 3 表示每秒发送 3 条消息。第 4 个参数 5 表示每条消息包含 5 个单词(实际上就是 5 个整数)。

执行完上面命令后，屏幕上会不断滚动出现新的单词，具体如下。

```
3 3 6 3 4
9 4 0 8 1
0 3 3 9 3
0 8 4 0 9
8 7 2 9 5
2 6 4 8 5
0 9 6 0 9
4 0 0 8 1
1 8 3 7 4
4 0 6 5 7
3 9 1 5 0
9 3 9 6 7
1 8 7 4 3
9 5 6 2 6
4 8 8 6 8
0 0 3 3 7
```

该终端窗口保持现状，千万不要关闭它，让它一直不断发送单词。之后我们新打开一个终端，执行以下命令，运行 KafkaWordCount 程序来执行词频统计。

```
$cd /usr/local/spark
$/usr/local/spark/bin/spark-submit--class "KafkaWordCount"
$/usr/local/spark/mycode/kafka/target/scala-2.12/simple - project_2.12-1.0.jar
```

运行完以上命令后，词频统计功能就被启动了，屏幕上就会显示以下信息。

```
SLF4J:Class path contains multiple SLF4J bindings.
SLF4J:Found binding in [jar:file:/usr/local/spark/lib/kafka/slf4j-log4j1.2-1.7.
21.jar!/org/slf4j/impl/StaticLoggerBinder.class ]
SLF4T: Found binding in [jar:file:/usr/local/hadoop/share/hadoop/common/
lib/slf4j-log4j12-1.7.10.jar!/org/slf4j/impl/staticLoggerBinder.claass ]
SLF4J: See http:// www.×××.org/codes.html # multiple _ bindings for
```

```
anexplanation.
SLF4J:Actual binding is of type [org.slf4j .impl. Log4jLoggerFactory]
-------------------------------------------
Time: 1479789000000 ms
-------------------------------------------
(4,16)
(8,14)
(6,15)
(0,10)
(2,9)
(7,17)
(5,14)
(9,9)
(3,8)
(1,8)
```

这些信息说明，Spark Streaming 程序顺利接收到了 Kafka 发来的单词信息，并进行词频统计且已得到结果。

11.5 本章小结

日志采集工具 Flume 和 Kafka 经常用在 Hadoop 和 Spark 生态系统中，用来进行日志信息的实时采集。

首先，本章介绍了日志采集工具 Flume 和 Kafka 的安装和使用方法，并给出了几个实例演示工具的具体用法。

然后，本章介绍了 Sqoop 的安装和使用方法。

最后，本章详细介绍了如何编写 Spark Streaming 应用程序来处理 Kafka 的数据源。通过这个实例，读者可以从总体上了解 Spark 和 Kafka 等工具之间的组合使用方法。